ADVANCES IN BIOCHEMICAL ENGINEERING

Volume 14

Managing Editor: A. Fiechter

With 39 Figures

Springer-Verlag
Berlin Heidelberg New York 1980

ISBN 3-540-09621-3 Springer-Verlag Berlin Heidelberg New York
ISBN 0-387-09621-3 Springer-Verlag New York Heidelberg Berlin

Library of Congress Catalog Card Number 72-152360
Printed in Germany

Typesetting, printing, and bookbinding: Brühlsche Universitätsdruckerei Gießen
2152/3140-543210

Managing Editor

Editorial Board

Contents

Bioconversions of Nitriles and Their Applications

Jean-Claude Jallageas
Laboratoire de Chimie Organique
Université des Sciences et Techniques du Languedoc
F-34060 Montpellier Cedex, France

Alain Arnaud and Pierre Galzy
Chaire de Génétique et Microbiologie
Ecole Nationale Supérieure Agronomique
F-34060 Montpellier Cedex, France

The bioconversions of nitriles and primary amides have a practical interest for the production of optically active α-hydroxy – or α-amino acids and for the preparation of highly pure amides and acids. The appropriate chemical hydrolyses are generally not suitable for such syntheses.

The literature concerning the reactions of nitrile catabolism by living organisms is analyzed. Among these reactions, only hydrolyses are involved in industrial processes.

A large number of economically important products may be obtained by hydrolyzing nitriles or primary amides: acrylamide, lysergic acid, DL- or L-lactic acid, DL- or L-alanine, DL- or L-methionine, DL- or L-phenylalanine, and D- or L-phenylglycine, to name a few. The descriptions of the principal known processes and the possibilities for their improvement are presented and discussed.

1 Introduction

A number of economically important organic compounds are industrially produced from nitriles by chemical synthesis (Table 1). The main processes are based on three chemical properties of nitriles:

- hydrogenation to amines

$$RCN \xrightarrow{+2\,H_2} RCH_2NH_2$$

- acid or basic hydrolysis to amides or organic acids resp.

$$RCN \xrightarrow{+H_2O} RCONH_2 \xrightarrow{+H_2O} RCOOH + NH_3$$

- action of bicarbonates on α-aminonitriles (reaction of Bücherer-Berg).

The production of amines from nitriles involves a rather complex methodology due to the presence of hydrogen under consuming pressure appreciable amounts of energy.

All reactions based on hydrolysis of nitriles or the Bücherer-Berg-reaction share some common disadvantages: formation of large quantities of salt, critical separation and isolation resp. of reaction products and prolonged heating. In addition, α-hydroxylic acids or optically active α-amino acids are never obtained.

Catalytic reactions represent a useful tool for hydratation of nitriles replacing acids or alkaly and simplifying technology by reducing the energy demand. The catalytical hydrolysis by the copper salts of acrylonitrile to acrylamide is the sole example of an industrial process of this type. Nevertheless, some side reactions are taking place in these reactions and the proper regeneration of the catalyst is difficult. On the other hand, laboratory experiments have been accomplished for optimation of the Strecker reaction.

Nitriles may be biologically transformed in order to avoid some of the disadvantages associated with all these chemical processes. The bioconversions occur under milder conditions, at pH values close to neutrality and at moderate temperatures. They also can lead to the synthesis of optically active α-hydroxy - and α-amino acids. In general biological reactions are of lower economy compared to chemical processes. Generally, they represent the only selection for the production of optically active compounds.

Considering the interesting possibilities offered by the bioconversions of nitriles, it is appropriate to examine the literature which deals with living organisms containing nitriles and which describes these natural compounds. These data are summarized in Table 2, where it may be seen that there are in fact relatively few known natural nitriles. It is true that published reports have preferentially treated organisms which liberate

Table 1. Industrial chemical processes leading to amides or acids from nitriles

Nitrile	Reaction	Product	Application of product
Adiponitrile $NC(CH_2)_4CN$	Pressure hydrogenation, 150 °C	Hexamethylene-diamine	Polymers
Nitriles of fatty acids (stearic acid, oleic acid, palmitic acid, lauric acid)	Pressure hydrogenation, 150 °C	Amines of fatty acids	Antiagglomerants, surfactants, emulsifiers, detergents, flotation agents
Acrylonitrile $CH_2 = CHCN$	Acid hydrolysis, hot	Acrylamide	Polyacryl-amide
Phenylacetonitrile $\Phi\ CH_2CN$	Acid hydrolysis, hot	Phenylacetic acid	Perfumes
Lactonitrile $CH_3-CH(CN)(OH)$ DL	Acid hydrolysis, hot	DL-Lactic acid	Therapeutics, dyes, tanning, lacquers
Mandelonitrile $\Phi\ CH(CN)(OH)$ DL	Acid hydrolysis, hot	DL-Mandelic acid	Antiseptics
α-Hydroxymethylthio-butyronitrile, DL $CH_3SCH_2CH_2CH(CN)(OH)$	Acid hydrolysis, hot	α-Hydroxymethyl-thiobutyric acid DL or MHA	Chicken feed
Acetone cyanohydrin $(CH_3)_2C(CN)(OH)$	Acid hydrolysis, hot	α-Hydroxyisobutyric acid then methacrylic acid	Plastics
α-Amino nitriles $RCH(CN)(NH_2)$	Acid hydrolysis or acid/basic hydrolysis, hot (Strecker reaction)	α-Amino acids	Animal feed, therapeutics, cosmetics, surfactants, polymers
β-Aminoproprionitrile $NH_2CH_2CH_2CN$	Acid/basic hydrolysis, hot	β-Alanine	Precursor of panthotenic acid
Malonitrile $CH_2(CN)(CN)$	Basic hydrolysis, hot	Malonic acid	Medicaments
α-Amino nitriles $R-CH(CN)(NH_2)$	Bücherer-Berg reaction	α-Amino acids	

Table 2. Natural nitriles

	Formula	Name	Organism[a] [Ref.]
Bacteria	NH_2, CN, N, N, N, ribose (ring structure)	Antibiotic 1037 or E-212 or naritheracin or toyocamycin or unamycin B or vengicide	*Streptomyces*[1–4]
Fungi	$R_1R_2C(CN)(O\ glucose)$	Cyanoglucosides	*Basidiomycetes*[5–6]
	$R_1R_2C(CN)(OH)$	Cyanhydrins	*Basidiomycetes*[7]
	$R{-}CH(CN)(NH_2)$	α-Aminonitriles	Basidiomycete W_2[8,9] *Rhizoctonia solani*[10]
	$R{-}C{\equiv}C{-}CN$	Acetylenic nitriles	*Clitocybe diataeta*[11–13] *Lepista diemii*[14] *Lepista glaucona*[15]
	O←N=N–CN on benzene ring with COOH	*p*-Carboxyphenylazoxycyanide	*Calvatia lilacina*[16, 17]
Algae	$R_1R_2C(CN)(O\ glucose)$	Cyanoglucosides	*Chlorella*[5, 6]
Plants	$R_1R_2C(CN)(O\ glucose)$	Cyanoglucosides	800–1000 species representing 70–80 families[5, 6]
	OH, CH_2CN, HO on benzene ring	2,4-Dihydroxyphenylacetonitrile	*Erica scoparia*[18]
	$(CH_3)_2CHCN$	Isobutyronitrile	*Theobroma cacao*[19]

Table 2 (continued)

	Formula	Name	Organism[a] [Ref.]
Plants	$\Phi\ CH_2CH_2CN$	3–Phenyl–propionitrile	*Nasturtium officinale*[20, 21]
	$\Phi\ CH_2CN$	Phenylaceto-nitrile	*Theobroma cacao*[19] *Tropaeolum majus, Lepidium sativum.*[20–22] *Leptactina senegambica, Codonocarpus cotinifolia.*[23]
	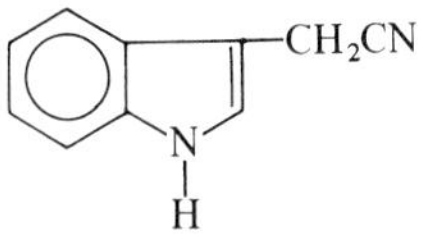 CH_2CN; N; H	Indoylaceto-nitrile	*Cruciferae, Lycopersicum esculentum,* etc.[24–29]
	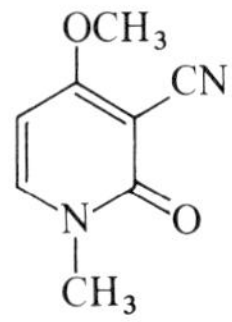 OCH_3; CN; N; O; CH_3	Ricine	*Ricinus communis*[30]
	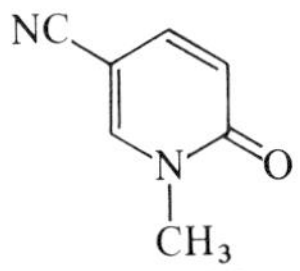 NC; N; O; CH_3	Nudiflorine	*Trewia nudiflora*[30–32]
	$NCCH_2CH(COOH)(NH_2)$	β-cyanoalanine	Leguminosae[33]
	$NCCH_2CH(COOH)(NHCOCH_2CH_2CH(NH_2)(COOH))$	N(γ-L-glutamyl)-β-cyanoalanine	Leguminosae[33]
	$NCCH_2 \cdot CH_2 \cdot NHCO(CH_2)_2CH(COOH)(NH_2)$	N(γ-L-glutamyl)-β-aminopropionitrile	*Lathyrus pusillus, Lathyrus odoratus*[34–36]
	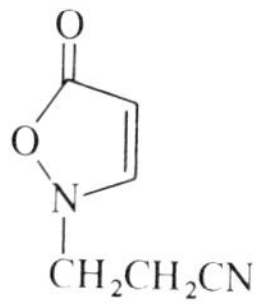 O; O; N; CH_2CH_2CN	2–(2–Cyanoethyl)–3–isoxalin–5–one	*Lathyrus odoratus*[36]

Table 2 (continued)

	Formula		Name	Organism[a] [Ref.]
	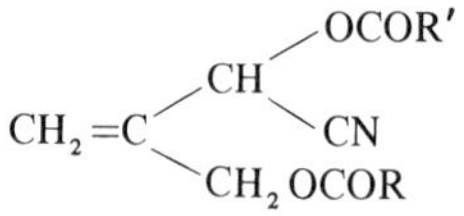 $CH_2=C(CH(OCOR')CN)CH_2OCOR$	I	Cyanolipids	Sapindaceae[38–40]
	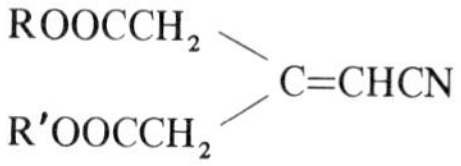 $(ROOCCH_2)(R'OOCCH_2)C=CHCN$	II		
	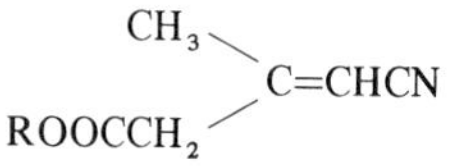 $(CH_3)(ROOCCH_2)C=CHCN$	III		
	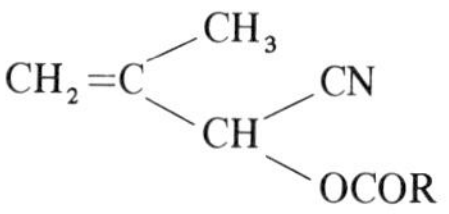 $CH_2=C(CH_3)CH(CN)OCOR$	IV		
	(with R, R' saturated or unsaturated, 14–22 carbons)			
Insects (anthropods)	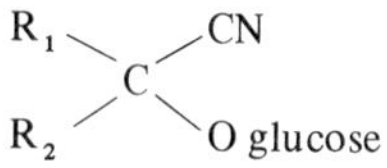 $R_1R_2C(CN)$O glucose		Cyanoglucosides	*Polydesmus vicinus, Paropsis atomoria.*[5, 6, 41–45]
	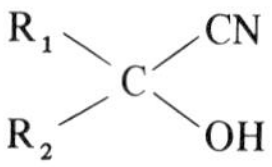 $R_1R_2C(CN)OH$		Cyanhydrins	*Diplopoda, Alpheloria corrugata, Harpaphe haydeniana*[46–53]
	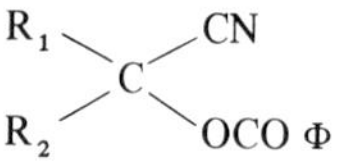 $R_1R_2C(CN)OCO\Phi$		Cyanhydrin benzoate	*Polydesmus collaris*[46]
Sponges	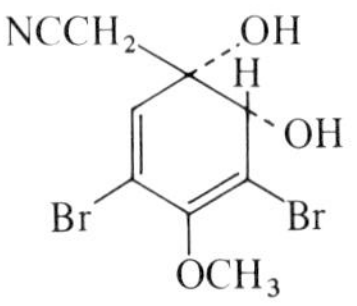 $NCCH_2$, OH, H, OH, Br, Br, OCH_3 (cyclohexadiene ring)		Aerophlysinin	*Lanthella, Verongia*[54]

[a] Species containing the cited product or the group to which the species belongs

HCN (cyanogenic organisms), a phenomenon found primarily in the plant kingdom. Furthermore, nitriles are difficult to isolate and identify. It is thus probable that a large number of natural nitriles exist, awaiting discovery. Table 2 also shows that nitriles are found in various families of organisms, except mammals for which they are toxic.

Natural cyano compounds and toxic nitriles must be catabolized by organisms and degraded to conventional products which are not toxic to cellular metabolism. The present review on the bioconversions of nitriles will thus initially treat their catabolism and biodegradation, particularly hydrolysis reactions. We will then consider the isolation and identification of bacterial strains with nitrilase activity. The possibilities of using these strains for bioconverting nitriles to organic acids and amides will be discussed. After a description of the uses of nitrilase-containing bacteria for the production of optically active α-amino acids, we will finally analyze the various biological processes for the preparation of acids from nitriles and primary amides described in the literature.

2 Catabolism and Biodegradation of Nitriles

2.1 α-Hydroxylation of Nitriles

Nitriles may be oxidized to cyanhydrins by oxygenases:

$$\begin{matrix} R_1 \\ R_2 \end{matrix} \rangle CH{-}CN \xrightarrow[\text{oxygenase}]{[O]} \begin{matrix} R_1 \\ R_2 \end{matrix} \rangle C \langle \begin{matrix} OH \\ CN \end{matrix}$$

This enzymatic pathway appears to be present in a large variety of organisms, including plants, fungi, insects, algae, sponges, and mammals. In spite of the fact that the corresponding enzymes and the intermediate α-hydroxynitriles have never been isolated, several proofs have been advanced for the existence of this reaction.

Some nitriles of the type RCH_2CN are transformed by plants to carboxylic acids RCOOH[55-58]. The most probable explanation for the synthesis of an acid with a shorter carbon chain is an initial α-oxidation of the nitrile, followed by the decomposition of the cyanhydrin to an aldehyde, which is then oxidized to an acid:

$$\left[RCH \langle \begin{matrix} CN \\ OH \end{matrix} \right] \rightarrow \begin{matrix} RCHO + HCN \\ \downarrow \\ RCOOH \end{matrix}$$

This hypothesis is supported by the finding that the metabolism of 3-indolylacetonitrile by wheat tissue involves the formation of 3-indolealdehyde[55].

In addition, it is probable that nitrile toxicity to mammals and insects results from the formation of HCN after α-hydroxylation[59-61]. Indeed:

– the administration of nitriles, as that of cyanide, causes the excretion of nontoxic thiocyanate in the urine; this thiocyanate arises from the action of CN^- on thiosulfate in the presence of rhodanese[62];

– a study of the action of a mouse liver extract on benzyl cyanide $C_6H_5CH_2CN$ *in vitro* demonstrated the synthesis of benzaldehyde C_6H_5CHO, which must have arisen after the α-hydroxylation of the nitrile[59];

– in the special case of ω-fluorinated nitriles, it has been shown that nitriles with an odd number of carbon atoms $2n + 1C$ are much more toxic than those with an even number $2nC$[63, 64]. This difference can be understood only by an α-hydroxylation of the nitrile, followed by an oxidation of the ω-fluorinated aldehyde to the corresponding acid and then by a β-oxidation degradation of the ω-fluorinated acid to acetic acid and to toxic fluoracetic acid (for $2n + 1C$ nitriles), or to nontoxic fluoropropionic acid (for $2nC$ nitriles).

Finally, the mechanism generally admitted for the biosynthesis of cyanoglucosides and cyanhydrins involves an intermediate α-oxidation of the nitriles obtained from α-amino acids[65–69]. This hypothesis is supported by the finding that HCN liberation by a large number of organisms very often depends on the presence of α-amino acids[70, 71].

2.2 Formation of Aldehydes from Cyanhydrins

Cyanhydrins give rise to ketone group and cyanhydric acid, either spontaneously or after the action of specific enzymes, oxynitrilases or hydroxynitrilelyases[72–80]. The enzymatic degradation of cyanhydrins has been demonstrated by UV-measurements due to a ketonic function in presence and absence of the enzyme.

$$R_1R_2C(CN)(OH) \rightleftharpoons R_1R_2C=O + HCN$$

This reaction is found in fungi, plants, and certain insects (antropods). The sorghum hydroxynitrilelyase has been extensively studied, particularly concerning its isolation and purification[72].

Natural cyanhydrins arise from the α-hydroxylation of nitriles (cf. 2.1) or from the enzymatic hydrolysis of cyanoglucosides, which liberates glucose and an aglycone moiety[81].

$$R_1R_2C(\text{O-glucose})(CN) \xrightarrow{\beta\text{-glucosidase}} R_1R_2C(OH)(CN) + \text{glucose}$$

2.3 Reduction of Nitriles

The nitrogenase present in algae and bacteria[82–85] is catalyzing the reduction of numerous important types of substrates like nitrogen, alcynes, allenes, cyanides, nitriles, isonitriles, cyanogen, azides, N_2O, H^+. Certain nitriles are similarly transformed to hydrocarbons (Table 3) releasing ammonia and probably also forming intermediately primary amines. Nitrogenase is found in procaryotes like bacteria, blue-green algae and actinomycetes considered as less developed organisms. It is possible that these types of organisms were involved in the formation of hydrocarbons from plant material. It is noteworthy that acetonitrile has been found in the lighter fractions of far distillates[86].

Table 3. Reduction of nitriles by nitrogenase

Nitriles	Products formed
$R-CN$	$R-CH_3 + NH_3$ ($R = CH_3, C_2H_5, C_3H_7$)
$CH_2 = CH-CN$	$CH_3CH=CH_2 + CH_3-CH_2-CH_3 + NH_3$
$CH_3-CH = CH-CN$ (*cis*)	$CH_3-CH_2-CH=CH_2 + CH_3-CH_2-CH_2-CH_3$ $+ CH_3-CH=CH-CH_3$ (cis) $+ NH_3$
$CH_3-CH = CH-CN$ (*trans*)	$CH_3-CH=CH-CH_3$ (trans) $+ NH_3$

Besides reduction of nitriles by nitrogenase the hypothetical enzymatic reduction of β-cyanoalanine to α,γ-diaminobutyric acid from *Lathyrus odoratus* has been never confirmed[87]

$$NCCH_2CH(COOH)NH_2 \rightarrow NH_2CH_2CH_2CH(COOH)NH_2$$

2.4 Hydrolysis of Nitriles

This reaction is the most common nitrile transformation. Thus, it is normal that different authors have attempted to demonstrate it in numerous and varied organisms. The nitriles tested belong to highly divers chemical types. Nevertheless the bibliographic data can be classed on the basis of the chemical type of the starting product and on the pathway employed beginning with the nitrile.

Nitrile hydrolysis may form amides with an arrest of the reaction at this point. There is a relatively limited number of examples of this type of reaction:

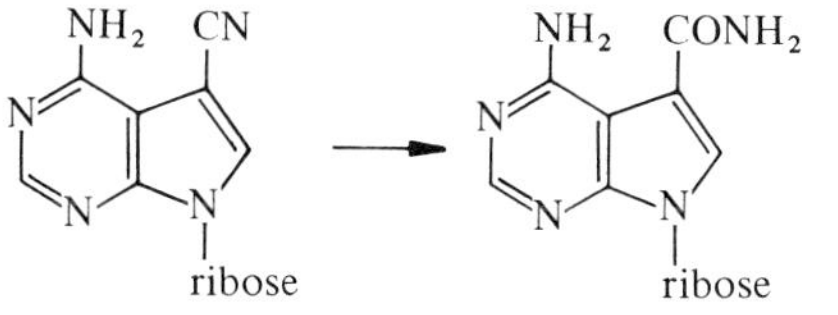

NH₂ CN — ribose → NH₂ CONH₂ — ribose

toyocamycin → sangivamycin

The transformation of a cyanopyrrolopyrimidine nucleoside has been described only in *Streptomyces rimosus*[88]. It is nonetheless probable that it also exists in other *Streptomyces sp.* which produce toyocamycin and sangivamycin.

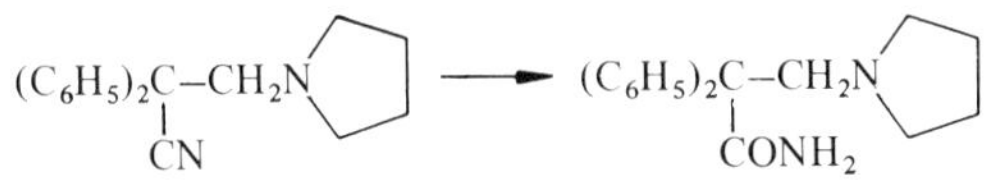

$$(C_6H_5)_2C(CN)-CH_2N(C_4H_8) \longrightarrow (C_6H_5)_2C(CONH_2)-CH_2N(C_4H_8)$$

2,2-diphenyl-3(1-pyrrolidino)-propionitrile → 2,2-diphenyl-3(1-pyrrolidino)-propionamide

The bioconversion of 2,2-diphenyl-3(1-pyrrolidino)-propionitrile found in *Penicillium*[89] must also be present in other organisms. Indeed, it has been reported that about 300 molds, basidiomycetes, and actinomycetes are capable of transforming this nitrilated substrate.

$$NCCH_2CH(COOH)(NH_2) \rightarrow NH_2COCH_2CH(COOH)(NH_2)$$

β-cyanoalanine → asparagine

This hydrolysis reaction has been described in numerous plants such as wheat, sorghum, etc.[90–92]. β-cyanoalanine is a natural nitrile found primarily in *Leguminosae*[33].

Nitrile hydrolysis can also give rise to acids without passing through the amide stage. This is particularly the case of the nitrilase described by Thimann and Mahadevan[93–95] and found in a certain number of plants. This enzyme converts 3-indolylacetonitrile and analogous compounds to the corresponding acids; it is not possible to demonstrate the presence of the corresponding amide at any moment of the reaction. The authors advanced the following reaction mechanism:

$$R-C{\equiv}N \xrightarrow{ESH} \begin{bmatrix} R-C{=}NH \\ | \\ SE \end{bmatrix} \xrightarrow{H_2O} \begin{bmatrix} R-C{=}O \\ | \\ SE \end{bmatrix} + NH_3$$

$$\begin{bmatrix} R-C{=}O & \\ |\nwarrow & \\ SE \quad {}^{-}OH & \\ \uparrow & \\ H^{+} & \end{bmatrix} \rightarrow RCOOH + ESH$$

ESH represents the enzyme, since essential sulfhydryl groups were demonstrated at the active site(s).

Direct hydrolysis of benzonitrile and of other aromatic nitriles has been described by Harper[96–98] in a *Fusarium* and a *Nocardia.* It was not possible to show the intermediate formation of an amide. In these two studies, the nitriles hydrolyzed had aromatic rings and thus the enzymes could have been particular nitrilases, specific for aromatic compounds.

In most cases, nitriles are hydrolyzed to acids with the formation of an amide intermediate. Literature references to this type of transformation appear in Table 4.

Table 4. Nitrile hydrolyses with amides as intermediates

Nitriles	Formulas	Organisms [Ref.]
Aliphatic Nitriles		
Acetonitrile	CH_3CN	*Corynebacterium nitrilophilus, Corynebacterium sp.*[99], *Corynebacterium pseudodiphteriticum*[100], *Pseudomonas sp.*[101], *Nocardia rhodochrous*[102]
Propionitrile	CH_3CH_2CN	*Nocardia rhodochrous*[102]
n-Butyronitrile, adiponitrile, butene-1-nitrile	$CH_3(CH_2)_2CN$, $NC(CH_2)_4CN$, $CH_2{=}CHCH_2CN$	*Corynebacterium pseudodiphteriticum*[100]
a-Hydroxynitrile		
Lactonitrile	$CH_3CH(CN)(OH)$	*Corynebacterium pseudodiphteriticum*[100]
Aromatic Nitriles		
Ricinine and analogs	3-cyano-2-pyridone ring, R_1 on N, R_2 at C-4	*Pseudomonas sp.*[103, 104]
Benzonitrile	Φ CN	*Corynebacterium pseudodiphteriticum*[100]
Dichlobenil	2,6-dichlorobenzene ring with CN (Cl, Cl, CN)	*Trichoderma sp., Penicillium sp., Fusarium sp., Geotrichum sp.*[105], *Soil bacteria*[106]
Bromoxynil	HO–benzene ring–CN with two Br	*Flexibacterium sp.*[107]

Table 4 (continued)

Nitriles	Formulas	Organisms [Ref.]
α-Amino Nitriles		
α-Aminopropionitrile	$CH_3CH(CN)NH_2$	*Corynebacterium sp.*[108]
α-Aminoisovaleronitrile	$(CH_3)_2CHCH(CN)NH_2$	*Corynebacterium sp.*[108]
Diverse		
β-Cyanoalanine	$NCCH_2CH(COOH)NH_2$	*Escherichia coli*[109–111], *Neurospora crassa, Lathyrus sylvestris, Lathyrus odoratus, Vicia villosa,* Guinea pig[112]
Indoylacetonitrile	indole-3-CH_2CN	Various plants[55, 113, 114]
2,4-Dichlorophenoxy-acetonitrile	2,4-$Cl_2C_6H_3OCH_2CN$	Various plants[55, 113, 114]
α-Cyano-3-phenoxy-benzyl-2,2,3,3-tetra-methylcyclopropane-carboxylate	2,2,3,3-tetramethylcyclopropane–CO–O–CHCN–C_6H_4–O–Φ	Undetermined soil organism[115]

Finally, in considerable research on the hydrolysis of nitriles to acids, no studies on the demonstration of an amide intermediate were performed. This is especially true in work on the toxicity of aromatic nitriles and aminoacetonitrile to mammals[116–118] and in research on waste water treatment by activated sludge[119]. In addition, no mention of passage through an amide intermediate was made in studies of the mechanism of DL-cyanhydrins hydrolysis to α-hydroxyacids by *Torulopsis candida* and of DL-α-amino-nitriles hydrolysis to α-amino acids by fungi[8–10].

2.5 Possibilities of Using Pathways of Nitrile Catabolism for their Bioconversions

The production of simple hydrocarbons by nitrogenase action on compounds as elaborate as nitriles is not economically interesting. The same is true for obtaining aldehydes from cyanhydrins.

Bioconversions using the other two nitrile catabolic pathways, however, seem to be promising. Biological α-hydroxylation could yield optically active cyanhydrins which could be hydrolyzed to yield the corresponding D- or L-α-hydroxyacids. The most economically interesting acid, L-lactic acid, could thus be synthesized from propionitrile:

$$CH_3CH_2CN \rightarrow \underset{(L)}{CH_3CHOHCN} \rightarrow \underset{(L)}{CH_3CHOHCOOH}$$

The above literature search also shows the numerous possibilities of applying biological hydrolyses of nitriles for the production of economically important amides and acids.

Relatively few amides are commercially available: acrylamide, benzamide, formamide, chloroacetamide, propionamide, salicylamide, nicotinamide, and phenylacetamide. Among them, only acrylamide is chemically obtained by nitrile hydrolysis; the others are prepared from acids or their derivatives by the action of ammonia. The corresponding nitriles are generally obtained by dehydration of amides and they are more expensive than the amides and acids. Thus, the only current important application of the biological hydrolysis of nitriles to amides is for acrylonitrile. This bioconversion could eliminate the production of secondary products present during chemical hydrolysis and reduce the energy requirement of the reaction.

There is a larger number of possibilities in the field of biological hydrolysis of nitriles to acids. Indeed, certain nitriles are much less expensive than the corresponding acids, e.g., α-amino acids, lactic acid, malonic acid, mandelic acid, methacrylic acid, phenylacetic acid, β-alanine, methylmercaptohydroxybutyric acid, and nitrilotriacetic acid. The biological production of these acids would have several advantages: avoid the production of a large quantity of salts, limit energy consumption and, above all, produce optically active acids.

We will now examine the research we have performed in an attempt to apply the biological hydrolyses of nitriles to the production of economically important amides and acids.

3 Study of Bacterial Strains with Nitrilase Activity

The above literature search showed that nitrile hydrolysis was found primarily in bacteria and fungi. Since the multiplication and utilization of bacteria are more manageable than those of fungi, it is logical to attempt the selection of bacterial strains with nitrilase activity.

3.1 Conditions of Isolation

Strains with nitrilase activity were selected from soil. They were isolated by plating a soil suspension in physiological saline on a medium containing 1.17% Yeast Carbon Base (Difco), 0.1% acetonitrile, and 2.5% agar. Following purification, each strain was tested in liquid medium containing acetonitrile as the only nitrogen source. Control inoculations were in the same medium minus acetonitrile. Strains capable of growing during four or five transfers in acetonitrile-containing medium were retained.

Acetonitrile was chosen as the test substrate because of its stability and solubility in water. It is nevertheless a volatile compound, and certain precautions had to be taken during the preparation of solid and liquid culture media. In addition, when large volume cultures were grown, it was necessary to use special reactors, identical to those used with methanol, in order to avoid acetonitrile evaporation during growth. Under our particular aeration conditions, evaporation in Erlenmeyer flasks was relatively low.

3.2 Identification of Isolated Strains

The strains isolated with nitrilase activity were all Gram-positive and obligate aerobes. They belong to the genera *Bacillus* and *Bacteridium* according to Prevot[120, 121], and to *Micrococcus* and *Brevibacterium* according to Bergey[120, 121].

3.3 Study of Nitrilase Activity

The nitrilated compounds used to determine the spectrum of nitrilase activity were either commercially purchased or synthesized with conventional techniques[122, 123].

Hydrolysis of the nitriles tested was performed with whole cells after growth on a complete medium, termed YMPG: 3 g yeast extract, 3 g malt extract, 5 g bacto-peptone, and 10 g glucose per l. Cultures were in Erlenmeyer flasks filled to 0.1 of their volume and were agitated for 16 h at 28 °C (80 oscillations min^{-1}, 8 cm amplitude). The culture was centrifuged and the bacteria were resuspended in a 4% aqueous nitrile solution whose pH had been previously adjusted to 8.0. After a 4–5 h incubation, the hydrolysis supernatant was recovered by centrifugation and analyzed.

The compounds obtained during hydrolysis were identified with relatively simple techniques, such as proton magnetic resonance spectrometry and chromatography.

Proton magnetic resonance spectrometry was used each time the peaks of the nitrile, amide, and acid equivalent protons were well separated on the PMR spectrum. The percent distribution of the various products was determined by integration.

In other cases, particularly during tests of the hydrolysis of 3-aminopropionitrile and of α-aminonitriles other than α-aminopropionitrile, we employed thin layer chromatography to identify the amides and amino acids in the reaction medium. Kieselgel 60 F 254 (Merck) was used as support; this silica gel is normally used to separate ionic or highly polar substances, such as amino acids. Plates were 20 · 20 cm and gel thickness was 0.25 mm. Two solvent systems were used: butanol: acetic acid: water (80: 20: 20, v/v/v) or propanol: 20% NH_4OH (70: 30, v/v). The first solvent enabled us to demonstrate the disappearance of the nitrile, while the second assured a better separation of the amide and acid. The developer was 0.2% ninhydrin in ethanol: acetic acid (80: 20, v/v). Thin layer chromatography was also used when assaying the hydrolyses of nicotinonitrile and isonicotinonitrile. In these cases, the solvent was 96% ethanol: chloroform: 20% NH_4OH (70: 40: 25, v/v/v). Spots were visualized with an ultraviolet lamp emitting at 254 nm.

The use of gas chromatography enabled us to demonstrate the disappearance of a large number of nitriles and the appearance of their hydrolysis products, amides and acids, characterized by their retention times. The column was filled with Porapak Q (80–100 mesh) with the following temperature conditions: detector, 300 °C; injector, 250 °C; oven, 130–250 °C, as a function of the compound. Organic acids could be analyzed only in acid solution; in order to avoid ghost peaks, certain operating precautions must be taken in the qualitative and quantitative determination of these acids[124)]. Special techniques were used in some cases: assays of NH_4OH and cyanide with specific electrodes, quantitative cyanhydrin, and cyanide assays by colorimetry[125)].

The nitriles tested, as well as the analytical method used, are found in Table 5. Among the 18 bacterial strains isolated, we limited ourselves to 9 strains in the study of the nitrilase activity spectrum.

- one *Bacillus* strain: R 332
- two *Bacteridium* strains: R 340 and R 341
- one *Micrococcus* strain: A 111
- one motile *Brevibacterium* strain: B 222
- four nonmotile *Brevibacterium* strains of different origins: A 13, B 212, C 211, and R 312.

All the nitriles tested were hydrolyzed to some extent by the nine strains in our experimental conditions. The bacteria tested thus have very wide nitrilase spectra.

It should be noted that this study was performed with whole cells and not homogenates, since the use of whole cells is the most suitable for industrial bioconversions. For a certain number of nitriles we also verified that hydrolyses could occur with homogenates.

It appears probable that these strains contain a rather nonspecific nitrilase, considering the wide substrate spectrum. We thus undertook a definition of the properties of this enzyme in one strain, *Brevibacterium* R 312, using the simplest nitrile substrate, acetonitrile. The kinetics of acetonitrile hydrolysis can be followed easily with proton magnetic resonance spectrometry[126)] or with gas chromatography[127)]. Knowledge acquired on the properties of acetonitrilase will be useful during the optimization of nitrile hydrolysis and also for the eventual immobilization of the enzyme on a solid support. Nevertheless, it remains to be shown by biochemical or genetic methods that this acetonitrilase is indeed a nitrilase with general activity.

We will now summarize the most important results of the acetonitrilase study[128)]. When a crude cell-free homogenate is prepared by ultrasonic disruption of the bacteria, enzyme activity is found in the supernatant of a 180,000 *g* ultracentrifugation. Activity

Table 5. Nitrilase activity spectra of bacterial strains with nitrilase and amidase activities (PMR: proton magnetic resonance; GLC: gas liquid chromatography; TLC: thin layer chromatography; E: specific electrode)

Class of compound	Formulas		Analytical method
Mononitriles			
Aliphatic nitriles	RCN	R = CH_3, C_2H_5, $(CH_3)_2CH$, $(CH_3)_3C$	PMR, GLC
		R = C_3H_7, iC_3H_7, C_4H_9	GLC
α-Ethylenic nitriles	$CH_2{=}C(CN)R$	R = H, CH_3	GLC, PMR
	$CH_3CH{=}CHCN$		GLC
β-Ethylenic nitrile	$CH_2{=}CHCH_2CN$		GLC
Aromatic nitriles	ArCN	Ar = Φ, p-toluyl, homoveratryl	GLC
α-Aminonitriles	$RCH(CN)NH_2$	R = H, CH_3	TLC, PMR
		R = C_2H_5, C_3H_7, $(CH_3)_2CH$, Φ CH_2, $CH_3SCH_2CH_2$	TLC
	$(CH_3)_2C(CN)NH_2$		TLC, PMR
α-Amino-nitriles N-Substitued	$CH_3CH(CN)NHR$	R = CH_3, CHO	PMR
α-Hydroxy-nitriles	$RCH(CN)OH$	R = H, CH_3	PMR, E_{CN^-}
		R = $CH_3SCH_2CH_2$	E_{CN^-}, Extraction
	$(CH_3)_2C(CN)OH$		PMR
β-Amino-nitrile	$NH_2CH_2CH_2CN$		TLC
Heterocyclic nitriles	RCN	R = 3-pyridyl, 4-pyridyl, 2-piperidyl	TLC
Dinitriles			
Aliphatic dinitriles	$NC(CH_2)_nCN$	n = 1, 2, 4	GLC
α-Aminodinitrile	$NCCH_2CH_2CH(NH_2)CN$		TLC

Table 5 (continued)

Class of compound	Formulas	Analytical method
Hetero-cyclic dinitrile	2,6-dicyanopiperidine (NC–C₅H₉N(H)–CN ring structure: NC, N, H, CN)	TLC, PMR
Diverse		
Cyanamide	NH_2CN	GLC, E_{NH_3}
Cyanhydric acid	HCN	GLC, E_{CN^-}, E_{NH_3}

is precipitated by ammonium sulfate between 40 and 55% of saturation. The pH optimum of acetonitrilase is pH 7, with activity falling off to zero at pH 5 and 10 (Fig. 1). Optimal temperature is 35 °C, with little variation in activity between 30 and 40 °C (Fig. 2). Heat denaturation studies showed that the enzyme is relatively thermolabile: although stable at 30 °, it is rapidly denatured at 35 and 40 °C and after 30 min at these temperatures, there remain 80 and 27%, respectively, of the initial activity (Fig. 3).

Considerable activity is also lost after dialysis against water or phosphate buffer, which suggests either a cofactor requirement or the association of enzyme subunits for activity. Since EDTA has no effect on activity, divalent cations are apparently not required. Finally, studies of the influence of the medium and the physiological state of

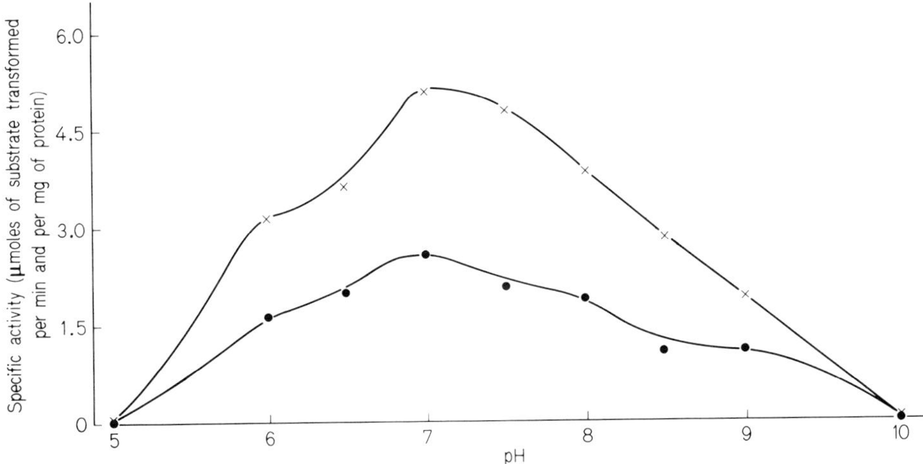

Fig. 1. Acetonitrilase activity as a function of pH at 30 °C
• Crude extract: supernatant of a 13,000 *g* centrifugation
x Supernatant of a 180,000 *g* ultracentrifugation

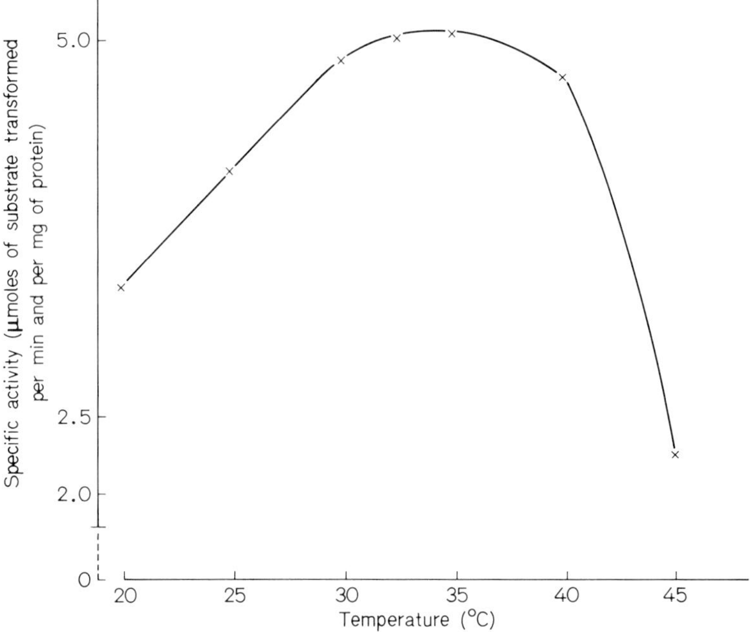

Fig. 2. Acetonitrilase activity as a function of temperature

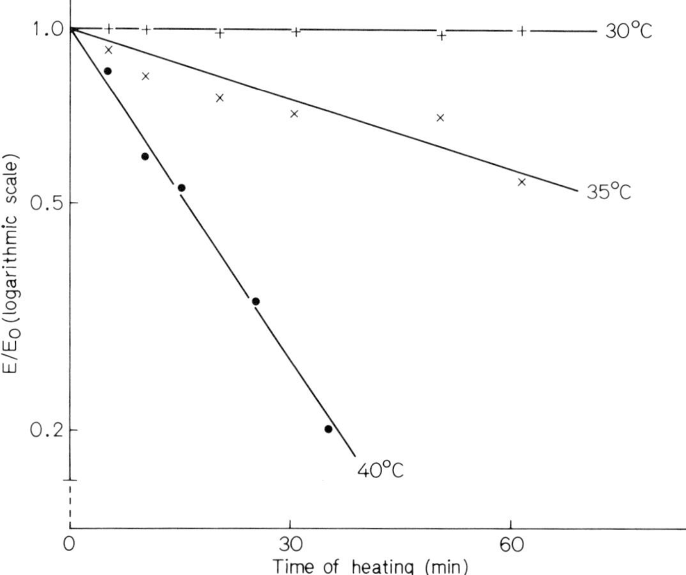

Fig. 3. Heat denaturation of acetonitrilase

E_0 = activity at time zero

E = activity after a given time of heating

The enzyme was suspended in 20 mM phosphate buffer, pH 7.3, and was heated at the temperatures indicated. The activity is determined at pH 7 and 30 °C

the cells showed that the enzyme is apparently constitutive. Enzyme activity decreases with decreasing growth rate and approaches zero at the stationary phase.

The Michaelis constant of 2.5×10^{-2} M is relatively high. This result favors the hypothesis according to which acetonitrilase is a general nitrilase, since nitrilases with very specific activities have much higher affinities for their substrates[88, 92, 94, 96, 104].

3.4 Study of Amidase Activity

The nine bacterial strains were next tested for their ability to hydrolyse caprolactam, α-amino-ε-caprolactam and the amides corresponding to the nitriles in Table 4, most of which were commercially available. When not available, the amides were synthesized by the action of gaseous HCl on the corresponding nitriles in formic acid[129]. α-amino-ε-caprolactam was obtained by the action of NH_4OH on α-chloro-ε-caprolactam[130, 131].

The conditions of culture (YMPG medium) and hydrolysis and the analytical techniques were the same as those used for the nitrilase studies. Thin layer chromatography was used for the qualitative analysis of caprolactam, α-amino-ε-caprolactam, and their hydrolysis products. The solvent system was *n*-butanol: water: acetic acid (6: 2: 2, w/w/w).

All the amides tested under the present conditions were hydrolyzed, except the vinylic amides (acrylamide, methacrylamide) and the lactams (caprolactam and α-amino-ε-caprolactam)[132]. The α-amino – and α-hydroxyacids obtained were racemic mixtures. Similar results were obtained when we used crude cell-free extracts.

The strains studied probably do not contain ε-lactamases and it is normal that an amidase, even one with nonspecific activity, would not attack these internal amides. It is more difficult to explain the lack of hydrolysis of the vinylic amides, since hydrolysis was tested with cells grown in Yeast Carbon Base medium containing 0.5% acetonitrile, acetamide or ammonium acetate as nitrogen source.

Amidase activity, similar to nitrilase activity, thus appears to be relatively nonspecific, and could be due to the action of an enzyme with little substrate specificity. We attempted to define certain properties of this enzyme, for the same reasons as in the case of the nitrilase. Acetamide was used as test substrate and quantitation of aqueous solutions was easily performed with proton magnetic resonance spectrometry[126] and gas chromatography[127].

The most important results[133] are as follows. The enzyme is localized in the supernatant of a 180,000 *g* centrifugation of the crude extract and is precipitated between 35 and 65% $(NH_4)_2SO_4$. The pH optimum is 7 and activity disappears at pH 3 and 11 (Fig. 4). Optimal temperature of the enzyme is included between 60 and 70 °C (Fig. 5). Acetamidase is relatively thermostable, losing only 25% of its activity after 3 h at 45 °C (Fig. 6). Activity is rapidly lost during dialysis. EDTA increases activity, probably by chelating an inhibitory divalent cation.

Acetamidase activity as a function of the physiological state of the cells and of the culture medium is shown in Table 6. In the exponential phase of growth, activity is 5–6 fold higher in cells growing in acetonitrile – or acetamide – supplemented medium than in cells grown with ammonium acetate; the enzyme is thus apparently inductive.

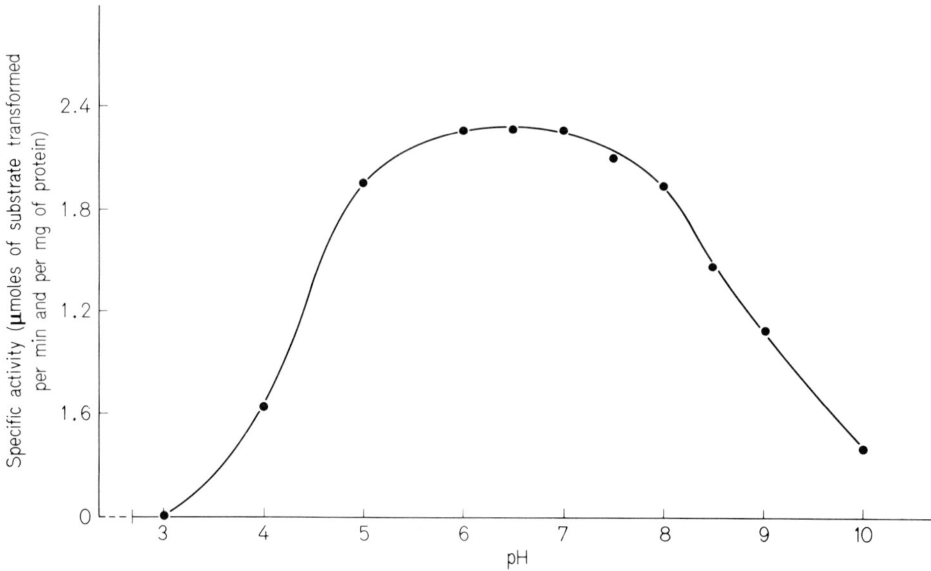

Fig. 4. Acetamidase activity as a function of pH
• S_2 : supernatant of a 180,000 *g* ultracentrifugation

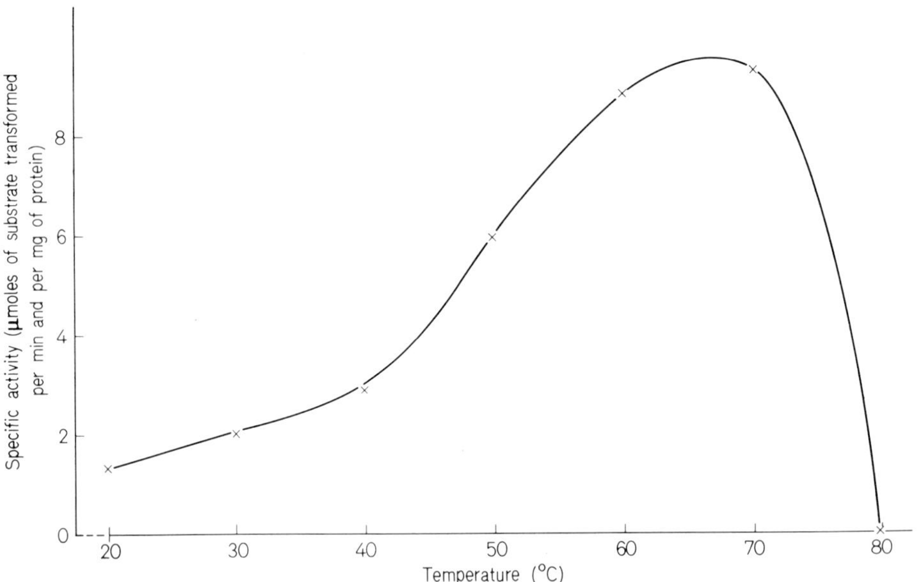

Fig. 5. Acetamidase activity as a function of temperature

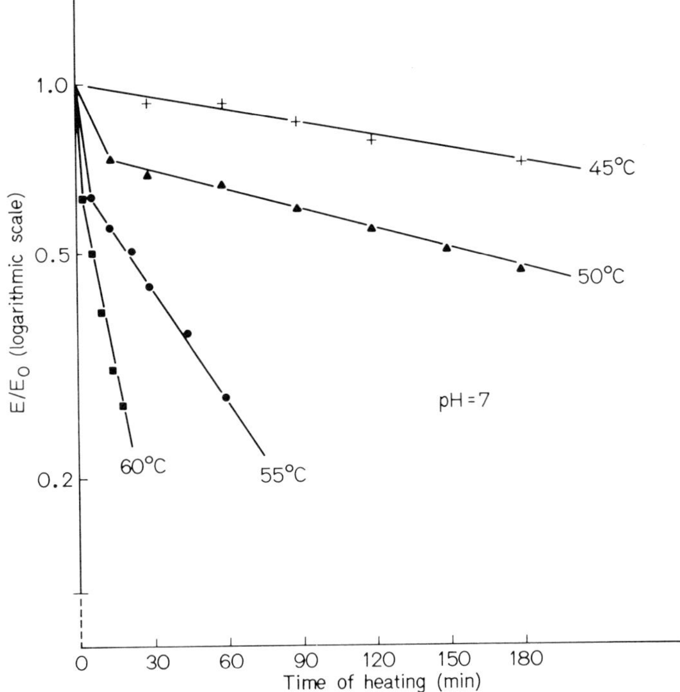

Fig. 6. Heat denaturation of acetamidase
E_0 = activity at time zero
E = activity after given time of heating
The enzyme was suspended in 20 mM phosphate buffer, pH 7.3, and was heated at the temperatures indicated. The activity is determined at pH 7 and 30 °C

Table 6. Effect of growth medium and physiological state of cells on acetamidase activity

Growth medium	Generation time (h)	Growth phase	Specific activity: μmoles of product transformed per min and per mg of protein at 30 °C and pH 7		
			Mean	Standard deviation	N
YCB + 0.5% acetonitrile pH 6.5	3	Exponential	1.90	0.30	5
		Stationary	1.55	0.20	3
YCB + 0.5% acetamide pH 6.5	3	Exponential	1.90	0.04	3
		Stationary	0.50	0.06	3
YCB + 0.5% NH_4 acetate pH 6.5	3	Exponential	0.35	0.03	3
		Stationary	0.60	0.06	3

In the stationary phase in YCB medium, activity is low in the presence of either acetamide or ammonium acetate and is high in YCB + acetonitrile. Gas chromatography demonstrated the absence of acetamide and the presence of acetic acid in YCB + acetamide and YCB + acetonitrile; in the latter medium, 20% of the initial acetonitrile had not been metabolized. The high amidase activity in stationary phase cells grown in acetonitrile medium is perhaps explained by this excess of acetonitrile, which could act as an enzyme inducer.

The Km of acetamidase is 3.5×10^{-3} M, similar to published values for aliphatic amidases[134, 135].

The results of this study show that bacterial acetonitrilase and acetamidase are considerably different, confirming that the hydrolysis of nitriles to acids in two steps occurs by the action of at least two different enzymes. It was thus consistent to search for acetamidase$^-$ mutants which have retained their acetonitrilase activity.

3.5 Search for an Acetamidase$^-$-Mutant

Strain R 312 was used for this study. The method used was adapted from Clarke and Tata[136] and involved selecting fluoracetamide-resistant mutants. This compound is toxic for wild-type, since it is transformed to fluoracetic acid by the action of the non-specific amidase. This screening thus enabled us to select mutants which could not perform this reaction.

Dense inocula of wild-type were plated on YNB, 0.5% ammonium acetate, 1% fluoracetamide, 1.2% agar, pH 6.5. Resistant colonies appeared after 8–10 d and were transferred to YNB, 0.5% ammonium acetate, pH 6.5, without fluoracetamide. Fluoracetamide resistance was then compared to wild-type in liquid medium. The mutant was then plated on YNB, 0.5% ammonium acetate, pH 6.5, to verify homogeneity and replica plating was then done on YNB, 0.5% ammonium acetate, 1% fluoracetamide, pH 6.5. Several defective mutants were thus screened, which were stable during vegetative division. One of these mutants, A_4, was used for the following studies.

The nitrilase activity spectrum in the mutant was identical to that of wild-type. This confirmed that there are two distinct enzyme systems involved in nitrile hydrolysis, nitrilase and amidase.

We then determined the amidase activity spectrum of this mutant and compared the activities of wild-type and mutant grown on YMPG medium. Hydrolysis conditions were the same as those used in the determination of the amidase spectrum of R 312 and cells were always tested in exponential phase. Activities were determined with whole cells suspended in water and with cell-free extracts obtained as previously described[137]. The results of this comparative study, as well as the analytical techniques used, are shown in Table 7.

Wild-type R 312 apparently hydrolyzes all the amides tested except acrylamide, methacrylamide, D- and L-asparagine, and D-glutamine, the latter of which is effectively hydrolyzed by the extract, implying that the compound does not penetrate from cells.

Table 7. Comparison of the acetamidase⁻ mutant and wild-type (PMR: proton magnetic resonance; TLC: thin layer chromatography; GLC: gas liquid chromatography; E_{NH_3}: NH_3-specific electrode)

Amides	Whole cells		Extract		Analytical method
	Wild-type	Mutant	Wild-type	Mutant	
$CH_3 \cdot CO \cdot NH_2$	+	–	+	–	PMR
$CH_3 \cdot CH_2 \cdot CONH_2$	+	–	+	–	PMR
$(CH_3)_2 \cdot CH \cdot CO \cdot NH_2$	+	–	+	–	PMR
$(CH_3)_3 \cdot C \cdot CO \cdot NH_2$	+	–	+	–	PMR
$CH_3 \cdot CHOH \cdot CO \cdot NH_2$	+	–	+	–	PMR
$(CH_3)_2 \cdot COH \cdot CO \cdot NH_2$	+	–	+	–	PMR
pyridine ring (N) with $CONH_2$ (3-position)	+	+	+	+	TLC
pyridine ring (N) with $CONH_2$ (4-position)	+	–	+	–	TLC
piperidine ring (N–H) with $CONH_2$ (2-position)	+	–	+	–	TLC
H_2NOC–piperidine ring (N–H)–$CONH_2$ (2,6-positions)	+	–	+	–	TLC
NH_2CONH_2	+	+	+	+	E_{NH_3}
$HCONH_2$	+	+	+	+	GLC-E_{NH_3}
$NH_2(CH_2)_2CONH_2$	+	–	+	–	TLC
$CH_3CH(CONH_2)(NHCHO)$	+	–	+	–	PMR
$CH_3CH(CONH_2)(NHCH_3)$	+	–	+	–	PMR
$NH_2CH_2CONH_2$	+	+	+	+	TLC
$CH_3CH(CONH_2)(NH_2)$ (D)	+	–	+	–	TLC-PMR
$CH_3CH(CONH_2)(NH_2)$ (L)	+	+	+	+	TLC–PMR

Table 7 (continued)

Amides	Whole cells		Extract		Analytical method
	Wild-type	Mutant	Wild-type	Mutant	
$CH_3CH(CONH_2)(NH_2)$ (DL)	+	+	+	+	TLC–PMR
$\Phi\ CH_2CH(CONH_2)(NH_2)$ (DL)	+	+	+	+	TLC
$C_2H_5CH(CONH_2)(NH_2)$ (DL)	+	+	+	+	TLC
$CH_3S(CH_2)_2CH(CONH_2)(NH_2)$ (DL)	+	+	+	+	TLC
$(CH_3)_2CHCH(CONH_2)(NH_2)$ (DL)	+	+	+	+	TLC
$(CH_3)_2CHCH_2CH(CONH_2)(NH_2)$ (DL)	+	+	+	+	TLC
$(C_2H_5)(CH_3)CH \cdot CH(CONH_2)(NH_2)$ (DL)	+	+	+	+	TLC
$NH_2 \cdot CO \cdot CH_2 \cdot CH(CO \cdot OH)(NH_2)$ (DL)	–	–	–	–	TLC
$NH_2 \cdot CO \cdot CH_2 \cdot CH(CO \cdot OH)(NH_2)$ (D)	–	–	–	–	TLC
$NH_2 \cdot CO \cdot (CH_2)_2 \cdot CH(CO \cdot OH)(NH_2)$ (L)	–	–	+	+	TLC
$NH_2 \cdot CO \cdot (CH_2)_2 \cdot CH(CO \cdot OH)(NH_2)$ (D)	–	–	–	–	TLC

Table 7 (continued)

$CH_2{=}CH \cdot CO \cdot NH_2$	–	–	–	–	GLC
$CH_2{=}C(CO \cdot NH_2)(CH_3)$	–	–	–	–	GLC
$CH_3 \cdot CH = CH \cdot CO \cdot NH_2$	+	–	+	–	GLC-E_{NH_3}
$CH_2{=}CH \cdot CH_2 \cdot CO \cdot NH_2$	+	–	+	–	GLC-E_{NH_3}
Φ $CO \cdot NH_2$	+	–	+	–	E_{NH_3}
$NH_2 \cdot CO \cdot CH_2 \cdot CO \cdot NH_2$	+	–	+	–	E_{NH_3}
$NH_2 \cdot CO \cdot (CH_2)_2 \cdot CO \cdot NH_2$	+	–	+	–	E_{NH_3}
$NH_2 \cdot CO \cdot (CH_2)_4 \cdot CO \cdot NH_2$	+	–	+	–	E_{NH_3}

Acetamidase activity of mutant A_4 is completely absent in extracts. In the presence of acetamide, whole cells yield traces of acetic acid, which appear to arise from another metabolic pathway. The only amides hydrolyzed were formamide, urea, nicotinamide, L-glutamine, and the α-aminoamides. It should be noted that whenever DL-α-aminoamides were used, hydrolysis never exceeded 50% and the α-amino acid isolated was the L-isomer. Additional results favor the stereospecificity of amidases: D-α-aminopropionamide is not hydrolyzed, while L-α-aminopropionamide is completely converted to L-α-alanine.

Wild-type *Brevibacterium* R 312 did not hydrolyze all the amides tested. L-glutamine is apparently not hydrolyzed as a result of a permeability barrier. Acetamidase activity in extracts was not completely nonspecific, but these results require more detailed studies; thus, it is interesting that acrylamide and methacrylamide are or are not hydrolyzed as a function of the culture medium. In any case, it remains probable that acetamidase has a relatively wide spectrum of activity, since the mutation we studied resulted in the loss of the ability to hydrolyze a large number of substrates. It is virtually impossible that a single mutation could have led to the loss of such a large number of enzyme species.

In addition to the relatively nonspecific enzyme termed acetamidase, *Brevibacterium* R 312 has several specific enzymes for the hydrolysis of urea, formamide, nicotinamide, glycinamide, L-glutamine, and L-α-amino acids.

In summary, we have at our disposal two types of strains:

1) wild-type R 312 can hydrolyze practically all water soluble nitriles to the corresponding acids, providing growth conditions are carefully chosen. These reactions are never stereospecific;

2) mutant A_4 converts nitriles to the corresponding amides (acrylamide, acetamide, lactamide, etc.) and can convert certain amides to the acids: nicotinamide, urea, formamide, glycinamide, and α-aminoamides. Among the latter, only the L-amide is hydrolyzed to the acid.

The above results concerning nitrilase – and amidase – containing bacteria will now be followed by the description of several technological applications of our strains. We will then analyze the patents and similar processes utilizing nitrile and amide bioconversions.

4 Examples of Technological Applications[137–139]

4.1 Production of DL-Lactic Acid

Wild-type R 312 was grown in a medium containing glucose as sole carbon source. Exponentially growing bacteria were centrifuged, washed with physiological saline and resuspended in a reaction medium consisting of 10% (w/v) chemically synthesized lactonitrile. The same concentration of lactonitrile may be obtained *in situ* by mixing equimolar aqueous solutions of acetaldehyde and HCN; the reaction is started by adjusting the pH to about 5 with concentrated ammonium hydroxide. In both cases the pH of the lactonitrile solution is adjusted to 7 before adding the bacteria. The quantity of cells added corresponded to about 20–40 g of dry matter l^{-1}; this resulted in the total hydrolysis of the nitrile in 2 – 3 h at 25 °C with constant agitation. Cells were then eliminated and the ammonium lactate in the supernatant could be quantitatively recovered by drying. It is also possible to quantitatively recover lactic acid by any known method, e.g., acidification followed by continuous extraction with diethyl ether or any other appropriate organic solvent.

It is important to note that the lifetime of the bacteria is relatively long under the above conditions and tests have shown that the same suspension may be used several times without loss of activity. Fixation of cells would probably improve their useful lifetime.

The same process may be used to quantitatively obtain various acids, e.g., DL-α-amino acids, DL-α-hydroxyacids, aliphatic acids, etc. Conversion is occasionally incomplete after 3 h, especially when using nitriles yielding only slightly water soluble amides (nicotinonitrile, adiponitrile) or when using β-aminopropionitrile. In the case of the latter the first step in hydrolysis is very rapid, but β-aminopropionamide hydrolysis occurs slowly. Yields may be substantially improved by the acquisition of constitutive mutants or of more detailed knowledge of the process.

4.2 Production of Acrylamide

As in the above section, mutant A_4 may be grown in any glucose-containing medium; the cells were centrifuged, washed with physiological saline and resuspended in reaction medium consisting of 6% (w/v) aqueous acrylonitrile, whose pH had been previously adjusted to 6.5 – 7 with KOH. The quantity of cells corresponded to 20–40 g of dry weight and total hydrolysis of the nitrile occurred in 20–30 min at 25 °C with constant agitation. The pH must be maintained constant at 6.5–7. After recuperating the supernatant by centrifugation, acrylamide may be recovered by chloroform extraction. If certain precautions are taken and if acrylonitrile is progressively added to the reaction, a final acrylamide yield of 20% (w/v) can be obtained.

As in the case of lactonitrile, it appears possible to improve cell utilization by fixation. Nitrilase immobilization on an appropriate support may also be an important technological improvement.

The procedure for preparing acrylamide may also be used to prepare other amides, such as acetamide, benzamide, succinamide, etc. The only industrial conversion which is presently interesting is that of acrylonitrile to acrylamide.

Certain other possibilities may be imagined in the future:

- the price of certain nitriles may decrease, thus making their bioconversion to the corresponding amides economically interesting;
- if new nitriles become commercially available, their bioconversion to amides may become important.

4.3 Production of L-Methionine

Mutant A_4 was grown on YMPG medium and exponentially growing cells were centrifuged, washed with physiological saline and resuspended in a reaction medium consisting of 6% (w/v) α-amino-γ-methylthiobutyronitrile hydrochloride, whose pH had been adjusted to 6.5 - 8.5 with KOH. The addition of 20–40 g dry weight of cells l^{-1} quantitatively converted the nitrile to 50% L-methionine and 50% D-methionine amide in 2–3 h. The reaction products were recovered by known techniques: crystallization of methionine at pH 7 after concentration of the supernatant or chromatography on an ion exchange resin.

The D-amide was converted to D-methionine by wild-type R 312 at pH 7.

A variation of this process involves the hydrolysis of α-aminonitrile previously synthesized in situ by reacting methylmercaptopropionaldehyde with NH_4Cl and KCN in 10 N NH_4OH at 40 °C for 1.5 h. Bacteria were added after adjusting the pH to 7.

Cell-free extracts may also be used to hydrolyze the nitrile to the L-α-amino acid. Furthermore, the method is not limited to methionine and we have successfully used it to prepare L-α-alanine, L-α-aminobutyric acid, L-phenylalanine, L-valine, and L-leucine. It is now possible to imagine the use of this method for the preparation of L-lysine, L-glutamic acid, L-tryptophan, and D- or L-phenylglycine from the corresponding α-aminonitriles already described in the literature. The yield of the process may also be improved by fixing the cells or by immobilizing the enzyme systems involved. Finally, a better knowledge of the conditions for inducing α-aminoamidases should enable the process to become competitive with other existing industrial processes. The L-α-aminoamidase activities presently considered are indeed inducible and the optimization of the medium is thus a prime goal.

5 Other Biological Hydrolyses of Nitriles and Primary Amides

5.1 Production of Lysergic Acid[140)]

This method involves the conversion of lysergamide or isolysergamide to lysergic acid with cultures of *Claviceps purpurea.* The reaction is performed in an aqueous solution of organic solvent (dimethylformamide or ethanol) for 5–10 days. The reaction product is extracted with n-butanol and purified. Although the process is relatively long, it enables one to obtain an important product under unusual conditions for the utilization of a protist.

This process may be usefully extended to other reactions which are difficult to realize in aqueous solution.

5.2 Production of L-α-Hydroxyacids[141)]

A strain of *Torulopsis candida* was inoculated into a medium containing 0.02% (w/v) racemic α-hydroxynitrile (C_3–C_7), inorganic salts, glucose and yeast extract. Cells were centrifuged away after 72 h of growth, the supernatant was brought to pH 10 and residual cyanhydrin was eliminated by ether extraction. The aqueous solution was then brought to pH 2 and treated with benzene, thus yielding 160 mg of L-α-hydroxyacid per l of culture; final yield was thus close to 75%.

Using this process, it is possible to obtain optically active L-α-hydroxyacids from racemic cyanhydrins, but the yields are currently low and the operating conditions are complex.

5.3 Preparation of Optically Active α-Amino Acids[142)]

The L- and DL-α-aminoamide precursors were dissolved in an aqueous medium pH 8.2, containing $MgCl_2$ and $MnCl_2$. After adding a crude cell-free extract of *Pseudomonas putida,* the reaction mixture was agitated for 20 h at 25 °C. L-α-aminoamide yields 100% of the corresponding L-α-amino acid and the DL-α-aminoamide yields 50% L-α-amino acid and 50% D-α-aminoamide. It is thus possible to imagine the production of L-α-amino acids from corresponding chemically synthesized DL-α-aminoamides. In addition, an α-aminoamide, D-phenylglycinamide may be racemized under certain conditions[143)]. This will result in the recycling of the residual D-amide and the acquisition of 100% L-phenylglycine from DL-phenylglycinamide.

This process for preparing optically active amino acids involves, as the procedure described above (Sect. 4.3), bacteria with L-α-aminoamidase activity. Nevertheless, our method has the advantage of utilizing racemic α-aminonitriles which are more easily synthesized than the corresponding α-aminoamides.

6 Conclusions and Perspectives

The various bioconversion processes that we have discussed have a number of interesting characteristics. There is low energy consumption, low salt concentrations, the absence of secondary products and a reduced evaporation of nitriles since we work at room temperature. In addition, these bioconversions may enable one to stereospecifically orient a reaction and thus produce optically active compounds, especially *α*-amino acids. There remain, however, a certain number of drawbacks related to current operating conditions. The quantity of cellular material used is relatively great and the yield of L-*α*-amino acids is, with the exception of phenylglycine, only 50% from racemic nitriles and *α*-aminoamides.

In spite of these limitations, we believe that certain improvements are possible, the most important of which are:

- the acquisition of inexpensive biological material by the improvement of culture conditions and a better control of the regulation of enzyme biosynthesis;
- the optimization of bioconversions by varying temperature, pH, activating cations, etc.;
- technological improvements, including cell fixation and enzyme immobilization. This improvement will enable us to continuously synthesize large quantities of amides and acids in a shorter time with the same quantities of cells or of enzymes;
- the development of a general method for racemizing D-*α*-aminoamides for the production of L-*α*-amino acids. This will enable us to obtain a quantitative yield of L-*α*-amino acid from the corresponding DL-*α*-aminonitrile.

The various processes considered and presented here were influenced by current market conditions. Certain of these bioconversions could of course be applied to other substrates (nitriles or amides) if they became commercially available at reasonable cost. It is also possible to use these processes for obtaining new products, especially relatively complex compounds.

7 Acknowledgments

The authors thank Mr. Brunie and Mr. Gillonnier of the "Laboratoire de l'Alimentation Equilibreé, Commentry, France (Groupe Rhône-Poulenc)" for synthesizing nitriles and amides.

8 References

1. Aszalos, A. et al.: J. Antibiot. (Tokyo) Ser. A *19*, 285 (1966)
2. Kikuchi, M.: J. Antibiot. (Tokyo) Ser. A *8*, 145 (1955)

3. Umezawa, H. et al.: Jpn. Patent *10*, 245 (1964)
4. Yamamoto, H. et al.: Ann. Rep. Takeda Res. Lab. *16*, 28 (1957)
5. Seigler, D.S.: Phytochemistry *14*, 9 (1975)
6. Eyjolfson, R.: Fortschr. Chem. Org. Naturst. *28*, 74 (1970)
7. Tapper, B.A., Mac Donald, M.A.: Can. J. Microbiol. *20*, 563 (1974)
8. Strobel, G.A.: J. Biol. Chem. *241*, 2618 (1966)
9. Strobel, G.A.: J. Biol. Chem. *242*, 3265 (1967)
10. Mundy, B.P., Liu, F.H.S., Strobel, G.A.: Can. J. Biochem. *51*, 1440 (1973)
11. Anchel, M.: J. Am. Chem. Soc. *74*, 1588 (1952)
12. Anchel, M.: Trans. N.Y. Acad. Sci. *16*, 337 (1954)
13. Anchel, M.: Science *121*, 607 (1955)
14. Thaller, V., Turner, J.L.: J. Chem. Soc. I, 2032 (1972)
15. Bohlmann, F., Arndt, C., Starnick, J.: Tetrahedron Lett. *24*, 1605 (1963)
16. Gasco, A. et al.: Tetrahedron Lett. *38*, 3431 (1974)
17. Bianco, M.A., Ceruti Scurti, J.: Allionia *18*, 79 (1972)
18. Ballester, A., Verwey, A., Overeem, J.C.: Phytochemistry *14*, 1667 (1975)
19. Van der Wal, B. et al.: Rec. Trav. Chim. Pays-Bas *87*, 238 (1968)
20. Hofmann, A.W.: Berichte *7*, 518 (1874)
21. Hofmann, A.W.: Berichte *7*, 1293 (1874)
22. Gadamer, J.: Berichte *32*, 2336 (1899)
23. Sabetay, S., Palfray, L., Trabaud, L.: C. R. Acad. Sci. Paris *207*, 540 (1938)
24. Nitsch, J.P., Nitsch, C.: Plant. Physiol. *30*, 55 (1955)
25. Jones, E.R.H. et al.: Nature *169*, 485 (1952)
26. Henbest, H.B., Jones, E.R.H., Smith, G.F.: J. Chem. Soc. *75*, 3796 (1953)
27. Denffer, D.V., Behrens, M., Fischer, A.: Naturwissenschaften *39*, 550 (1952)
28. Bennett-Clark, T.A., Kefford, N.P.: Nature *171*, 645 (1953)
29. Kutacek, V., Kefeli, V.I.: In: Biochemistry and physiology of plant growth substances. Wigtman, F., Setterfield, G. (eds.), p. 127. Ottawa: The Ring Press Ltd 1968
30. Marion, L.: In: The alcaloïds: chemistry and physiology, Vol. 1, p. 206. New York: Academic Press 1950
31. Mukherjee, R., Chatterjee, A.: Chem. Ind., 1524 (1964)
32. Mukherjee, R., Chatterjee, A.: Tetrahedron *22*, 1461 (1966)
33. Bell, E.A.: In: Chemotaxonomy of the leguminosae. Harbone, J.B., Boulter, D., Turner, B.L. (eds.), p. 179. London, New York: Academic Press 1971
34. Schilling, E.D., Strong, F.M.: J. Am. Chem. Soc. *77*, 2843 (1955)
35. Tschiersch, B.: Pharmazie *21*, 445 (1966)
36. Schilling, E.D., Strong, F.M.: J. Am. Chem. Soc. *76*, 2848 (1954)
37. Van Rompuy, L. et al.: Biochem. Biophys. Res. Commun. *56*, 199 (1974)
38. Mikolajczack, K.L.: Prog. Chem. *15*, 97 (1977)
39. Seigler, D.S.: Biochem. Syst. Ecol. *4*, 235 (1976)
40. Gowrikumar, G., Mani, V.V.S., Lakshminarayana, G.: Phytochemistry *15*, 1566 (1976)
41. Pallares, E.S.: Arch. Biochem. *9*, 105 (1946)
42. Moore, B.P.: J. Aust. Entomol. Soc. *6*, 36 (1967)
43. Blum, M.S., Woodring, J.P.: Science *138*, 512 (1962)
44. Jones, D.A., Parson, J., Rothschild, M.: Nature *193*, 52 (1962)
45. Barbetta, M., Casnati, G., Pavan, M.: Mem. Soc. Entomol. *45*, 169 (1966)
46. Casnati, G. et al.: Experientia *19*, 409 (1963)
47. Eisner, T.: Ann. Rev. Entomol. *7*, 107 (1962)
48. Jacobson, M.: Ann. Rev. Entomol. *11*, 403 (1966)
49. Duffey, S.S., Underhill, E.W., Towers, G.H.N.: Comp. Biochem. Physiol. B *47*, 753 (1974)
50. Eisner, T. et al.: Science *139*, 1218 (1963)
51. Eisner, H.E., Eisner, T., Hurst, J.J.: Chem. Ind. (London) *124* (1963)
52. Eisner, H.E., Alsop, D.W., Eisner, T.: Psyche *74*, 107 (1967)

53. Eisner, T., Eisner, H.E.: Natural History *74*, 30 (1965)
54. Fulmor, W. et al.: Tetrahedron Lett. *52*, 4451 (1970)
55. Fawcett, C.H. et al.: Proc. Roy. Soc. London B *148*, 543 (1958)
56. Fawcett, C.H. et al.: Nature *176*, 1026 (1955)
57. Taylor, H.F., Wain, R.L.: Nature *184*, 1142 (1959)
58. Chamberlain, K., Wain, R.L.: Ann. Appl. Biol. *75*, 409 (1973)
59. Ohkawa, H. et al.: Pesti. Biochem. Physiol. *2*, 95 (1972)
60. Frankenberg, L., Sorbo, B.: Arch. Toxikol. *31*, 99 (1973)
61. Contessa, A.R., Santi, R.: Biochem. Pharmacol. *22*, 827 (1973)
62. Lang, K.: Biochem. Z. *259*, 243 (1933)
63. Pattison, F.L.M.: Nature *172*, 1139 (1953)
64. Pattison, F.L.M. et al.: J. Am. Chem. Soc. *78*, 3484 (1956)
65. Kjaer, A., Lansen, P.O.: Biosynthesis *2*, 171 (1973)
66. Kjaer, A., Lansen, P.O.: Biosynthesis *4*, 179 (1976)
67. Conn, E.E.: Biochem. Soc. Symp. *38*, 277 (1974)
68. Schutte, H.R.: In: Fortschritte der Botanik, Vol. 35, p. 103. Berlin: Springer Verlag 1973
69. Hughes, M.A., Conn, E.E.: Phytochemistry *15*, 697 (1976)
70. Knowles, C.J.: Bacteriol. Rev. *40*, 652 (1976)
71. Gewitz, H.S. et al.: Planta *131*, 149 (1976)
72. Seely, M.K., Criddle, R.S., Conn, E.E.: J. Biol. Chem. *241*, 4457 (1966)
73. Mao, C.H., Anderson, L.: Phytochemistry *6*, 473 (1967)
74. Burckhardt, H., Kleinschmidt, T.: Z. Physiol. Chem. *348*, 753 (1967)
75. Kleinschmidt, T., Glossmann, H., Horst, J.: Z. Physiol. Chem. *351*, 349 (1970)
76. Colotelo, N., Ward, E.W.B.: Nature *189*, 242 (1961)
77. Bove, C., Conn, E.E.: J. Biol. Chem. *236*, 207 (1961)
78. Stevens, D.L., Strobel, G.A.: J. Bacteriol. *95*, 1094 (1968)
79. Haisman, D.R., Knight, D.J.: Biochem. J. *103*, 528 (1967)
80. Diericks, P., Wauters, E., Vendrig, J.: Z. Pflanzenphysiol. *75*, 191 (1975)
81. Butler, G.W., Bailey, R.W., Kennedy, L.D.: Phytochemistry *4*, 369 (1965)
82. Hardy, R.W.F., Burns, R.C., Pashall, G.W.: In: Inorganic biochemistry, Vol. 2, Chap. 23, p. 745. Eichkorn, G. (ed.). Amsterdam: Elsevier 1973
83. Postgate, J.R.: The chemistry and biochemistry of nitrogen fixation. London: Plenum Press 1971
84. Schrauzer, G.N. et al.: J. Am. Chem. Soc. *94*, 7378 (1972)
85. Fuchsman, W.H., Hardy, R.W.F.: Bioinorg. Chem. *1*, 197 (1971)
86. Vincent, C., Delachenal, C.: C. R. Acad. Sci. Paris *86*, 340 (1879)
87. Ressler, C., Redstone, P.A., Erenberg, R.H.: Science *134*, 188 (1961)
88. Uematsu, T., Suhadolnik, R.J.: Arch. Biochem. Biophys. *162*, 614 (1974)
89. Theriault, R.J., Longfield, T.H., Zaugg, H.E.: Biochemistry *11*, 385 (1972)
90. Fowden, L., Bell, E.A.: Nature *206*, 110 (1965)
91. Castric, P.A., Conn, E.E.: J. Bacteriol. *108*, 132 (1971)
92. Castric, P.A., Farnden, K.J.F., Conn, E.E.: Arch. Biochem. Biophys. *152*, 62 (1972)
93. Thimann, K.V., Mahadevan, S.: Nature *181*, 1466 (1958)
94. Thimann, K.V., Mahadevan, S.: Arch. Biochem. Biophys. *105*, 133 (1964)
95. Mahadevan, S., Thimann, K.V.: Arch. Biochem. Biophys. *107*, 62 (1964)
96. Harper, D.B.: Biochem. Soc. Trans. *4*, 502 (1976)
97. Harper, D.B.: Biochem. J. *165*, 309 (1977)
98. Harper, D.B.: Biochem. J. *167*, 685 (1977)
99. Mimura, A., Kawano, T., Yamaga, K.: J. Ferment. Technol. *47*, 631 (1969)
100. Grant, D.J.W.: Antonie van Leeuwenhoek *39*, 273 (1973)
101. Firmin, J.L., Gray, D.O.: Biochem. J. *158*, 223 (1976)
102. Di Geronimo, M.J., Antoine, A.D.: Appl. Environm. Microbiol. *31*, 900 (1976)
103. Robinson, W.G., Hook, R.H.: J. Biol. Chem. *239*, 4257 (1964)

104. Hook, R.H., Robinson, W.G.: J. Biol. Chem. *239*, 4263 (1964)
105. Pate, D.A., Funderburk, H.: Proz. Symp. Lise Isotopes Weed. Res. *12*, 17 (1965)
106. Fournier, J.C.: Chemosphere *3*, 77 (1974)
107. Smith, A.E., Cullimore, D.R.: Can. J. Microbiol. *20*, 773 (1974)
108. Fukuda, Y. et al.: J. Ferment. Technol. *49*, 1011 (1971)
109. Farnden, K.J.F.: Ph. D. Thesis, University of Adelaïde (1970)
110. Lauinger, C., Ressler, C.: Biochem. Biophys. Acta *198*, 316 (1970)
111. Jackson, R.C., Handschumacher, R.E.: Biochemistry *9*, 3585 (1970)
112. Giza, Y.H., Ratzkin, H., Ressler, C.: Fed. Proc. *22*, 651 (1963)
113. Stowe, B.B., Thimann, K.V.: Arch. Biochem. Biophys. *51*, 499 (1954)
114. Seeley, R.C. et al.: Chemistry and mode of action of plant growth substances. London: Butterworths Scientific Publications 1956
115. Roberts, T.R., Standen, M.E.: Pestic. Sci. *8*, 600 (1977)
116. Williams, R.T.: In: Detoxication mechanisms. New York: John Wiley and Sons 1959
117. Bray, H.G., Hybs, Z., Thorpe, W.V.: Biochem. J. *48*, 192 (1951)
118. Ponseti, I.V. et al.: Proc. Soc. Exp. Biol. Med. *93*, 515 (1956)
119. Ludzack, F.J. et al.: Sewage Ind. Wastes *31*, 33 (1959)
120. Arnaud, A., Galzy, P., Jallageas, J.C.: C. R. Acad. Sci. Paris *287*, 571 (1976)
121. Arnaud, A., Galzy, P., Jallageas, J.C.: Rev. Ferment. Ind. Aliment. *31*, 39 (1976)
122. Marvel, C.S., Brubaker, M.M.: Organic syntheses, Vol. 1. New York: John Wiley and Sons Inc. 1948
123. Bejaud, M. et al.: Tetrahedron *31*, 403 (1975)
124. Du Preez, J.C., Lategan, P.M.: J. Chromatogr. *124*, 63 (1976)
125. Bruce, R.B., Howard, J.W., Hanzal, R.F.: Anal. Chem. *27*, 1346 (1955)
126. Jallageas, J.C., Arnaud, A., Galzy, P.: Anal. Biochem. *95*, 436 (1979)
127. Jallageas, J.C., Arnaud, A., Galzy, P.: J. Chromatogr. *166*, 181 (1978)
128. Arnaud, A., Galzy, P., Jallageas, J.C.: Agric. Biol. Chem. *41*, 2183 (1977)
129. Becke, F., Fleig, H., Paessler, P.: Liebigs Ann. Chem. *749*, 198 (1971)
130. Wineman, R.J., Eu-Phang T. Hsu, Anagnostopoulos, C.E.: J. Am. Chem. Soc. *80*, 6233 (1958)
131. Francis, W.C. et al.: J. Am. Chem. Soc. *80*, 6238 (1958)
132. Arnaud, A., Galzy, P., Jallageas, J.C.: Folia Microbiol. *21*, 178 (1976)
133. Jallageas, J.C., Arnaud, A., Galzy, P.: J. Gen. Microbiol. *24*, 103 (1978)
134. Jakoby, W.B., Fredericks, J.: J. Biol. Chem. *239*, 1978 (1964)
135. Thalenfeld, B., Grossowicz, N.: J. Gen. Microbiol. *94*, 131 (1976)
136. Clarke, P.H., Tata, R.: J. Gen. Microbiol. *75*, 231 (1973)
137. Arnaud, A. et al.: CNRS – ANVAR Fr. 73,33613 (19 Décembre 1973)
138. Galzy, P. et al.: CNRS – ANVAR Fr. 74,41828 (18 Décembre 1974)
139. Jallageas, J.C., Arnaud, A., Galzy, P.: CNRS – ANVAR Fr. 79,01803 (24 Janvier 1979)
140. Societa Farmaceutici Italia: Belg. Pat. 634, 836 (January 13, 1964)
141. Kawaken Fine Chemicals Company: Jap. Pat. 9042886 (September 3, 1972)
142. Stamicarbon, B.V.: Lux. Pat. 74,142 (8 Janvier 1976)
143. Stamicarbon, B.V.: Pays-Bas Pat. 75,14301 (9 Décembre 1975)

Ergot Alkaloids and Their Biosynthesis

Zdeněk Řeháček
Institute of Microbiology
of the Czechoslovak Academy of Sciences
14220 Prague 4, Czechoslovakia

This work reviews recent ergot alkaloid (EA) literature and illustrates the present problems of EA research. Biosynthesis of EA is regulated genetically. Many enzymes and coenzymes of basic metabolism and EA synthesis are common. A polygenic system is assumed for the control of EA and EA phenotypes. Formation of EA is to a greater degree a result of changes in the physiological state than an indication of a new physiological state. EA are not waste, physiologically inert products of cell metabolism. The clavine alkaloid producers are biocybernetically similar. In contrast, the biocybernetics of the organisms forming peptide EA are much more specific. The EA future depends on the results of basic research, including the horizontal and vertical integration of various disciplines. The newer EA derivatives have opened up a new area of investigation.

1 Introduction

Progress in every experimental region depends on the development of methods which logically divide the studied subject up into smaller components. This approach is necessary but not sufficient. The divergence of the individual directions in research has made it impossible for a single individual to comprehend the whole of a given field. Thus it is important in each area to understand the stage of analytical development where it is possible and useful to commence at least partial synthesis. The combination of the analytical and synthetic approaches enables exact solution of even complex problems. This is particularly true for ergot alkaloids, the research into which, application and production depends on the results of cooperation of plant physiologists, microbiologists, biochemists, organic chemists, pharmacologists, and chemical technologists.

Ergot (Secale cornutum) – sclerotium of pyrenomycete *Claviceps purpurea* (Fr.) Tul. – has a long history. Over the centuries its role underwent important changes from a dreaded toxic parasite on rye to an important source of biologically effective substances. At the beginning of the fifties, the biosynthesis of ergot alkaloids drew a great deal of attention. It rapidly changed from a subject of fruitful speculation to the region of experimental research. Primarily the chemical properties of the alkaloids were studied. In the meantime, biologists were studying the physiological aspects of the synthesis of alkaloids from a practical point of view, i.e., for both agronomists, who were trying to obtain grain with the lowest possible alkaloid content and for pharmacognosists, who were, on the other hand, interested in high alkaloid yields. The dependence of the pharmacological effectiveness on the alkaloid chemical structure, enabling preparation of substances for specific purposes, was important stimulus for the development of ergot alkaloid research. The isolation of two enzymes catalyzing the formation of ergot alkaloids, i.e., dimethylallyltryptophan synthetase in 1971 and chanoclavine cyclase in 1973, was also an important stimulant. These facts and the range of practical applications of ergot alkaloids and their semisynthetic derivatives, either alone or in composite preparations, led to the carrying out of this work.

This work will not only be a review of ergot literature, but also will illustrate the present problems of ergot research, with emphasis on "white spots", considering aspects worth pursuing further and it should also stimulate integration of individual fields of alkaloid research.

2 Importance

The ergot alkaloids have many potential therapeutic uses, some of which have been realized and others that have not survived the critical test of time. The ergot alkaloids have been of prime interest to pharmacologists due to their multiple actions, which include effects on uterine and vascular smooth muscle, and for their use as a tool for studying the mechanism of the sympathetic nervous system. Within the last few years newer aspects of the pharmacology of the ergot derivatives have been recognized[1]. One involves the action of certain ergot derivatives as inhibitors of prolactin secretion. In addition, the newer ergot compounds have been shown to produce stimulation of dopaminergic receptors and have been utilized effectively in the treatment of parkinsonism. Thus, these newer ergot derivatives have been implicated as potential therapeutic agents in many disease states such as parkinsonism, acromegaly, amenorrhea-galactorrhea, suppression of postpartum lactation, treatment of breast cancer, and possibly cancer of the prostate (secondary to inhibition of prolactin secretion).

3 Sources

In nature, ergot alkaloids are formed primarily by the fungus *Claviceps.* More than 40 different alkaloids have been isolated from this pyrenomycete, which has about 50 species. Ergot alkaloids have also been found in other fungi[2], especially of the *Aspergillus*[3], and *Penicillium*[4] genera and even in some higher plants of the *Convolvulaceae*-family[3].

Ergot alkaloids can be obtained in three basic ways:

1. By isolation from the pyrenomycete *Claviceps purpurea* prepared from field production (95% of world production);

2. by extraction of saprophytic cultures of various ascomycetes, especially of the *Claviceps* genus (Table 1);

3. by partial or total synthesis.

Table 1. Industrial submerged production of ergot alkaloids[5]

Alkaloid	Company[a]			
	1	2	3	4
Ergocornine	+	+	+	+
Ergocristine	+	+		+
Ergocryptine	+	+	+	+
Ergotamine		+	+	
Ergometrine			+	
Lysergic acid		+		
Paspalic acid	+			+

[a] 1 Biochemie GmbH, Kundl/Tirol, Austria
2 Farmitalia S.p.A., Milano, Italy
3 Gedeon Richter, Budapest, Hungary
4 Sandoz-Wander AG, Basel, Switzerland

Partial alkaloid synthesis is employed in studying of the structure and pharmacological effect of peptide alkaloids with various peptide parts of the molecule. The total synthesis of lysergic acid, the solution of the spacial arrangement of this substance, the synthesis of the peptide parts of some alkaloids and especially the total synthesis of ergotamine, enabling the synthesis of most ergot alkaloids with very important results[6]. However, total synthesis remains uneconomical.

4 Chemistry

The chemistry of ergot alkaloids has been reviewed in several works[3, 7, 8, 1]. A common part of ergot alkaloids is ergoline, i.e., partially hydrogenated indolo(4,3-fg)quino-

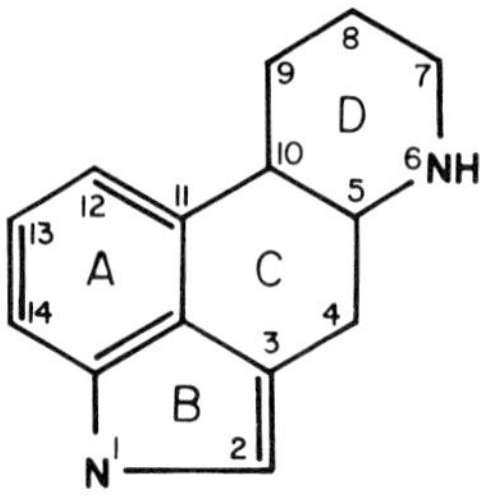

Fig. 1. Ergoline[188]

line (Fig. 1). Most natural ergot alkaloids form 6,8-dimethyl-$\Delta^{8,9}$ or $\Delta^{9,10}$-ergolene derivatives. Ergot alkaloids can be separated into two basic groups: derivatives of lysergic acid and clavine alkaloids[9]. Bu' Lock and Barr[10] also include tricyclic chanoclavines in the clavine group. The derivatives of lysergic acid so far isolated are the amides, in which the amide part is a small peptide or a simple alkylamide. The basicity of the amides of lysergic acid is a result of the presence of nitrogen in the 6-position (Fig. 1). The classic peptide ergot alkaloids, i.e., ergopeptines, are characterized by a modified tripeptide containing L-proline[11] and α-hydroxy-α-amino acid, which formed cyclol with a carbon from the proline carboxylic group[12–15]. Ergot alkaloids are the only natural substances with demonstrated cyclol structure. Alkaloids with hydrogens in positions 5 and 10 which are spatially alternately oriented are the subject of considerable pharmacological interest. Relatively few new natural alkaloids have been described in the last ten years. Most of them have only biogenetic importance. The discovery of ergostine and of ergostinine[16], as representatives of a new, i.e., ergoxine, group of peptide alkaloids is important.

4.1 Molecular Flexibility of Ergopeptines

It is clear from the molecular formulae of ergopeptines (Fig. 2) that the A/B/C- and the E/F-ring system each forms a rigid molecular fragment. Apart from these fixed fragments, the molecules have in principle a fairly large number of degrees of conformational freedom, mainly rotations about formal single bonds, such as C(8)–C(18), N(20)–C(2′), C(5′)–C(16′), and C(16)–C(17′), together with pseudorotations in the two saturated rings D and G. From crystallographic studies of ergopeptines one may infer that much of this apparent flexibility is limited by various intramolecular forces[17]. Prominent among these internal interactions is a strong hydrogen bond between cyclol hydroxyl O(13′) and carboxyl oxygen O(19). Together with the invariably observed trans peptide bond C(18)–N(20), this hydrogen bond fixes the torsion angles around the single bond N(20)–C(2′) within a narrow range.

The proline ring G shows minor conformational variations in the different structures. The flexible benzyl group H, however, appears to be restricted to a certain orientation in four of the investigated structures, while another orientation is adopted in two other structures. Steric hindrance between neighboring atoms is probably responsible for this reduced flexibility. The D-ring conformation is, however, the major structural determi-

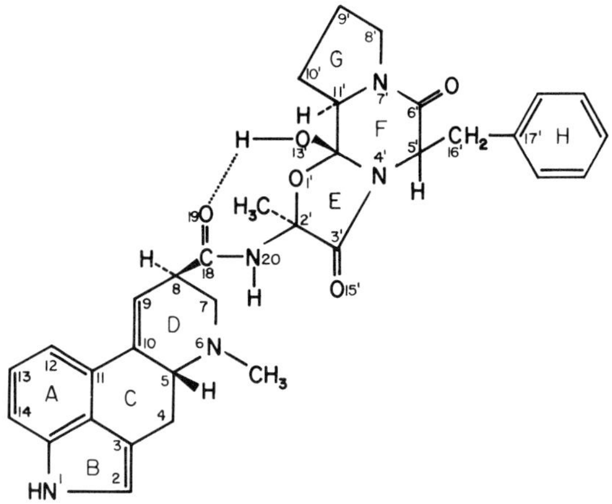

Fig. 2. Molecule of ergotamine[1)]

nant for various ergopeptines. In the biologically active compounds ergotamine and dihydroergotamine the amide group is equatorial to the D-ring, while in the biologically inactive forms this group is axial and stabilized in this position by an internal hydrogen bond N(20)–H...N(6) (Fig. 3). The torsion angles around C(8)–C(18) in the structures of ergotamine and dihydroergotamine fall in a rather narrow range, such that the plane of the amide group is roughly normal to the mean plane of the D-ring. This restricted rotation may be attributed to van der Waals contacts between the atoms adjacent to the C(8)–C(18) bond.

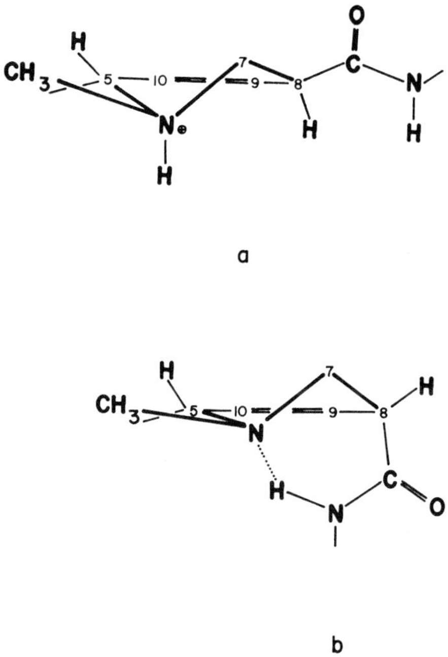

Fig. 3. The D-ring partial conformation of derivatives of lysergic (**a**) and isolysergic (**b**) acids

4.2 Chemical Manipulation

The ergot alkaloids can be chemically manipulated to form derivatives that possess altered pharmacologic effects. The selective saturation of one of the double bonds (C 9 to C 10) of the lysergic acid moiety results in compounds which differ considerably from the natural alkaloids[18]. One of the most prominent divergencies was found in respect to the uterine effects. The dihydrogenated alkaloids lost the excitatory effect of the uterus, but they are indeed able to inhibit in vitro and in situ the powerful stimulative action of the natural alkaloids such as ergotamine and ergotoxine[19].

5 Formation Physiology

Most *Claviceps* strains form alkaloids under conditions of parasitic growth. After the first proof of the formation of ergot alkaloids in vitro[20], a number of works were devoted to saprophytic strains of *Claviceps*. Only some of them synthesized alkaloids. Almost 1000 different species of filamentous fungi were tested for their ability to produce alkaloids under surface and submerged cultivation conditions[21–23]. The ability to synthesize clavine alkaloids under laboratory conditions was demonstrated for the *Claviceps* strains, for *Aspergillus fumigatus*, and for 19 further strains of phycomycetes, ascomycetes and imperfect fungi.

While metabolic pathways taking part in the synthesis of ergot alkaloids have been studied in detail, physiological formation of alkaloids has been studied only negligibly. The formation physiology does not consist only in culture growth and synthesis of the ergoline skeleton from corresponding precursors[24]. The formation of alkaloids requires changes in the metabolic structure normally leading to propagation of the biomass, a decrease in the intensity of culture growth and a decrease in the number of nuclei in the terminal cells of hyphae. Biosynthesis of alkaloids is to a greater degree a result of changes in the physiological state than an indication of a new physiological state. It is accompanied by decreased propagation intensity, accumulation of primary metabolic intermediates and a change in differentiation[25]. Differentiation is considered to be a result of different expression of the genome under various conditions and at different times[190, 191]. Morphological differentiation is preceded by synthesis of membranes, i.e., proteins and lipids. Protein synthesis is quantitatively regulated during morphological changes[192]. For initiation of differentiation, macromolecular turnover with minimum requirements on synthesis of new macromolecules is important.

At the beginning of their differentiation, vegetative fungal cells form a polarization gradient in the cytoplasm. Respiration plays an important role in maintaining polarized growth. Hyphae growing under aerobic conditions require active mitochondria if they are to dominate in cultures over cells with yeast growth. In vegetatively polarized hyphae, apical vesicles move through the cytoplasm along an electrochemical gradient produced by the oxygen. These vesicles contain precursors of the cell walls and hydrolyzing or synthesizing enzymes. The amitochondrial peak of vegetative hypha is a place with high reduction ability provided glycolysis occurs during catabolite repression[26]. This marked reduction ability is a result of sulfhydryl groups, which are very numerous at the hypha end[26, 27]. The sulfhydryl groups have a causal relationship to active alcoholic glycolysis catalyzed by SH-enzymes, e.g., hexokinase, glyceraldehyde hydrogenase and alcohol dehydrogenase. In the direction from the hypha end to the region rich in oxidative mitochondria there is a redox or electrochemical gradient. In catabolic derepression, in which exogenic glucose is exhausted, a vegetative polarization gradient disappears, mitochondria enter the hypha end and provide it with the metabolic energy necessary for sporulation syntheses. Turian[28] feels that glycolytic activity has a "vegetating" role, as conversion of the metabolism to the oxidative one induces conidia formation.

Conidiation is a complicated process. In *Aspergillus nidulans* cultures conidia formation is controlled by 300–800 genes[29]. With *Claviceps purpurea,* the formation of conidia is inversely dependent on the synthesis of alkaloids for those saprophytic strains which partially retain parasitic development, i.e., differentiation of the sphacelial phase to the conidial and/or sclerotial phase. In contrast, in those strains for which the conidial phase, the phase of vegetative growth and the alkaloid phase are not clearly separated, conidia formation occurs simultaneously with alkaloid synthesis. These strains are, however, quite sensitive to external conditions[30]. Physiological conditions necessary for conidiation and germination of conidia are different from conditions controlling alkaloid synthesis[31]. The conidiation process is affected by similar factors as the formation of alkaloids, but is not connected directly with alkaloid formation. Sekiguchi and Gaucher[32] came to similar conclusions in a study of the strain *Penicillium urticae* producing patulin.

The producents of ergot alkaloids do not represent only the spatial organization of cells and their morphological characteristics, but also organization in time of processes which occur in the organism. The cells form ergot alkaloids in a relatively short time interval of their development. In submerged cultures, alkaloid synthesis occurs under conditions of decreased cell proliferation and clearly defined degradation and resynthesis of nucleic acids and proteins[33–35]. With increased alkaloid formation, the proteolytic activity of the mycelium decreases[36, 37]. Many authors[10, 38, 39] observed intense proteosynthesis during alkaloid formation, however without a subsequent increase in the overall content of cell proteins. These results indicate a substantial protein turnover[38]. Amino acids freed during protein degradation pass into the metabolic pool of the mycelium and take part in various synthetic processes. Řeháček et al.[40] demonstrated that there is a positive relationship between the synthesis of alkaloids and the activity of glutamine synthetase converting most of the free cell ammonia into nontoxic glutamine. Glutamine then takes part in the synthesis of tryptophan – the precursor of alkaloids and inducer of their formation.

Cultivation conditions suitable for metabolism of primary substances are not suitable for alkaloid metabolism[41]. Submerged *Claviceps* cultures, for example, form alkaloids after using up the phosphate in the medium and accumulating polyols, lipids, polyphosphates, nonstructural carbophosphates and also inducers and precursors of alkaloids. Successful course of the alkaloid synthesis requires the presence of a higher citrate level or some other intermediates in the tricarboxylic acid cycle in the cell. The producing cell adjusts the level of these intermediates by accepting them from the medium or by synthesis. Acceptance of citrate from the medium can be temporarily increased by increasing the concentration of phosphate in the medium (Pažoutová, unpublished). An indicator for the synthesis of alkaloids is the storage of polysaccharides with $\beta(1-4)$ and $\beta(1-6)$ bonds of subunits in the space between the cell wall and the cytoplasm[42, 43]. The production of ergot alkaloids is correlated positively with the synthesis of lipids[42] (Fig. 4). Results of study of the *Candida* strain which forms lipids[44] support the opinion that microorganisms which eccessively accumulate lipids lack a control mechanism. Consequently they lack, e.g., phosphofructokinase and are characterized by marked derepression of the enzymes of the ribulosophosphate pathway, which is the source producing $NADPH_2$ necessary for lipid formation.

Cells generally synthesize alkaloids on poorly assimilable carbon sources. It can be assumed that the formation of alkaloids occurs simultaneously with induction of enzymes which provide the cell with energy from sources other than glucose. Glucose and other rapidly metabolizable sources of carbon depress the formation of ergot alkaloids[45]. For alkaloid processes a high C:N ratio is necessary[46]. The C:N ratio does not change during the process (Pažoutová, unpublished). It must, however, be borne in mind that with the *Claviceps* strain, similarly as with other filamentous fungi, syntrophic phenomena are important. A further requirement for submerged alkaloid formation is a nutrient medium with a high osmotic pressure[47, 48]. The importance of the respiration chain and the presence of oxygen for the synthesis of alkaloids is documented by the results of the work of Mátlová and Řeháček[49]. Oligomycin, an inhibitor of oxidative phosphorylation, depresses the growth of submerged cultures of *Claviceps purpurea* and alkaloid synthesis. Antimycin A blocks electron transfer in the section between cytochrome b and cytochrome c_1, and induces opposite alternation of the intensity of culture growth and of alkaloid formation. These results also support the theory on the positive role of the clavine alkaloid cycle in the energy metabolism of the producing cells[50].

Decreased differentiation and decrease in cell proliferation are one of the reasons for morphological variability of submerged mycelium of *Claviceps*. The character of fat hyphae and "sclerotia" of submerged mycelium (Fig. 4 b) is reminiscent of the plectenchymatic structure of production sclerotia of parasitic cultures. The formation of sclerotial cells can be artificially induced by ethidium bromide (0.05 mM), acriflavine (0.05 mM) or fluorescein (1 mM). A similar effect can be attained by physiological manipulation of the cultivation conditions, e.g., by using media with citric acid, ammonium sulphate and an increased content of Ca^{2+} (Pažoutová et al., unpublished).

The region of formation of ergot alkaloids is not generally identical with the place where alkaloids are found in increased concentrations. The members of *Ipomoea* genus form alkaloids in leaves and accumulate them in seeds, which do not produce the alkaloids. In submerged *Claviceps* cultures the location of clavine alkaloid synthesis is the endoplasmic reticulum, which later forms vacuoles containing lipoproteins[51]. Clavines and simple derivatives of lysergic acid are found primarily in the medium. This cannot, however, be considered as an indication of active alkaloid transport. The ratio between

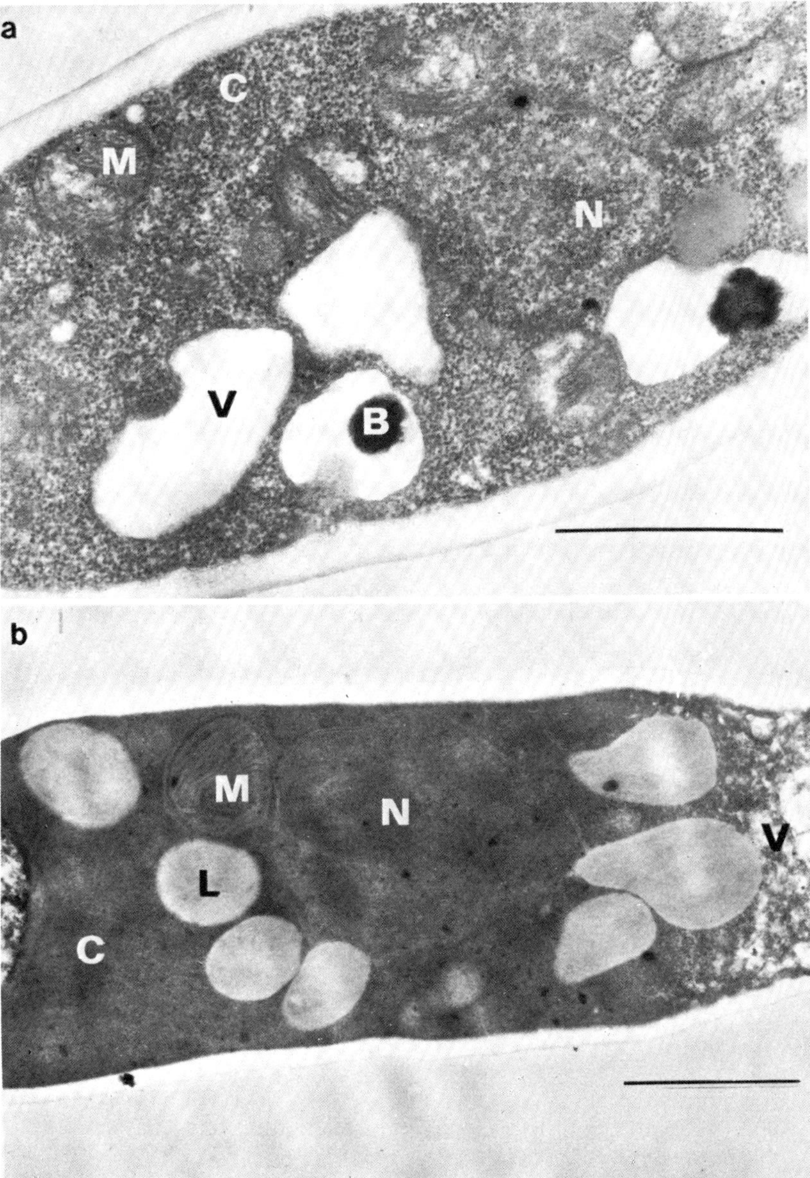

Fig. 4. Ultrastructure of the submerged mycelium of Claviceps purpurea[42)]

a) Sphacelial mycelium cell nonproducing alkaloids. C, cytoplasm; M, mitochondria; N, nucleus; V, vacuole; B, vacuolar body. Bar = 1 μm

b) Productive mycelium cell. C, cytoplasm; M, mitochondria; N, nucleus; L, lipid droplet; V, vesicle (vacuole). Bar = 1 μm

intra- and extracellular liquid volume is 1 : 25, so that 95% of the alkaloids are in the medium at equilibrium between alkaloid contents in the cell and in the medium. In contrast, peptide ergot alkaloids are mostly intracellular. It is probable that, in addition to accumulation in vacuoles, they also accumulate in the cell walls and plastides, which also contain negatively charged groups able to bind alkaloid cations.

It was long assumed that ergot alkaloids are final products which are not metabolized further. Elymoclavine *(Ipomoea rubrocoerulea)* is, however, degraded metabolically similar, e.g., to morphine *(Papaver somniferum)* and caffeine *(Coffea arabica)*. Robbers et al.[52] feel that enzymes degrading ergot alkaloids are activated by phosphate.

5.1 Biogenesis

Ergot alkaloids are produced under conditions of substrate limitation other than by carbon from key intermediates in basic metabolism by pathways differing from basic metabolism. Many enzymes and coenzymes of basic metabolism and alkaloid synthesis are common. According to cybernetics, alkaloids are formed, like other substances, in two ways at least (Fig. 5); otherwise, regulation of their formation would not be possible. The ergot alkaloid skeleton is derived from tryptophan (I), mevalonic acid (II) and methionine (VII)[53–56]. The treatment of these precursors depends on the biocybernetics of the given individual, i.e., on the appropriate genes being present in the cell, on their activation and deactivation in dependence on changes in the medium and also on structural changes. Genuine ergot alkaloids are levorotary.

The central precursor of the ergoline skeleton is tryptophan. The initial alkylation of tryptophan is an important step in the biosynthesis of alkaloids. The five-carbon unit derived from mevalonic acid (II) is incorporated as isopentenylpyrophosphate (IV) or the closely related hemiterpenoid in position C (4) of the decarboxylated tryptophan[57]. Alkylation of tryptophan in the unusual C(4) position is probably either caused by enzyme controlled steric and/or electronic changes in tryptophan or is a result of N- or O-prenylation of side-chain of tryptophan and subsequent intramolecular shift of an alkyl group to position C(4) by internal electrophilic substitution[58]. Heinstein et al.[59] and Lee et al.[60] prepared 4-dimethylallyltryptophan (V) from γ,γ-dimethylallylpyrophosphate and L-tryptophan using a mycelium extract of *Claviceps purpurea*. Petroski and Kelleher[61, 62] carried out the enzyme synthesis of 4-(E-4′-hydroxy-3′-methyl-but-2′-enyl)-L-tryptophan (HDMAT) (VI). They used isopentenylpyrophosphate (IV), L-tryptophan and a mycelium extract of *Claviceps paspali* for the preparation of HDMAT, which also appears in the medium during alkaloid biosynthesis[63]. Petroski and Kelleher[61] also demonstrated the incorporation of HDMAT into the lysergic acid amide molecule

---

Fig. 5. Some biogenetic relationships among ergot alkaloids[61, 74, 75, 94, 95, 133, 152, 181–187].
I tryptophan, II mevalonate, III dimethylallylpyrophosphate, IV isopentenylpyrophosphate, V 4-dimethylallyltryptophan, VI 4-(E-4′-hydroxy-3′methyl-but-2′-enyl)-L-tryptophan, VII methionine, VIII chanoclavine-I, IX hydroxychanoclavine-I, X agroclavine, XI elymoclavine, XII lysergol, XIII lysergene, XIV paspalic acid, XV lysergic acid, XVI lysergylalanine, XVII ergometrine, XVIII α-hydroxyethylamide of lysergic acid

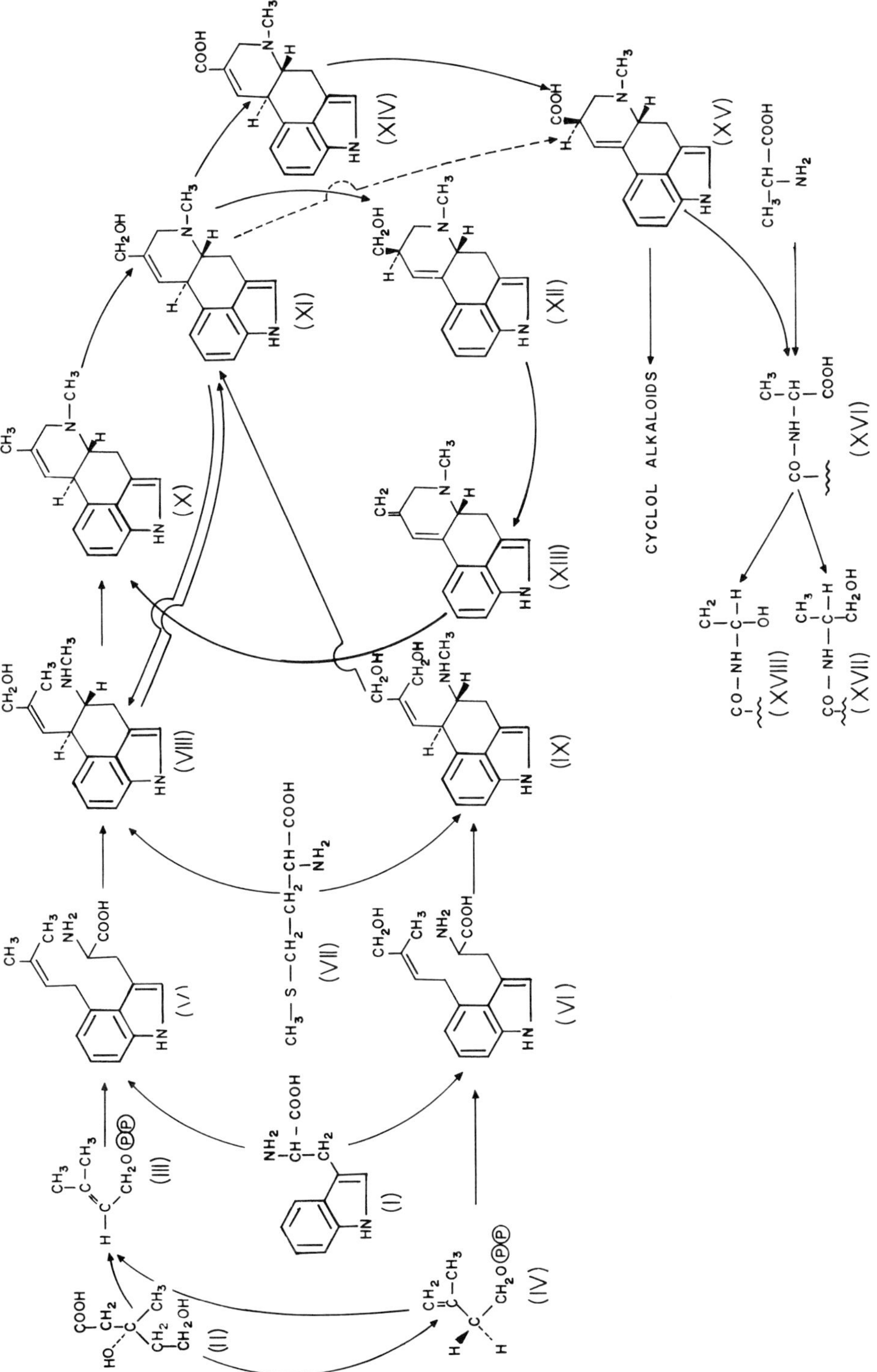
(I)
(II)
(III)
(IV)
(V)
(VI)
(VII)
(VIII)
(IX)
(X)
(XI)
(XII)
(XIII)
(XIV)
(XV)
(XVI)
(XVII)
(XVIII)
CYCLOL ALKALOIDS

and consider HDMAT as the precursor of the hydroxychanoclavine-I molecule (IX), from which elymoclavine (XI) is directly produced.

Mevalonic acid (II), which is another building block of ergolines, is neither the limiting substrate nor part of the regulation of the biogenesis of alkaloids[64–66]. Mevalonic acid may also be formed by the malonyl CoA pathway with participation of biotin dependent carboxylation[67]. Biotin takes part in the synthesis of mevalonic acid and ergot alkaloids by the fungus *Aspergillus fumigatus*[68].

In comparison with the other groups of alkaloids, the relationships among ergot alkaloids were studied in greater detail. Clavine alkaloids lie at the beginning of the biogenetic order and peptides of lysergic acid are at the end (Fig. 5, 6). The oxidation order agroclavine (X) → elymoclavine (XI) → lysergic acid (XV) is the main biogenetic pathway. Hydroxylation of the methyl group of agroclavine is a specific reaction of the *Claviceps* species[66] and requires the presence of cytochrome P-450[70]. In many fungi agroclavine and elymoclavine are 8-hydroxylated[71]. It is generally felt that the conversion of agroclavine to elymoclavine occurs through the presence of peroxidase or oxygen transferase. Jindra et al.[72] found that the oxygen of the hydroxyl group of elymoclavine originates from molecular oxygen. The precursor of the agroclavine molecule is chanoclavine-I (VIII) and not isochanoclavine-I[73, 74]. Hydroxychanoclavine-I (IX) is considered to be the precursor of the elymoclavine molecule (XI). The conversion of clavine into derivatives of lysergic acid passes through paspalic acid (XIV). In contrast

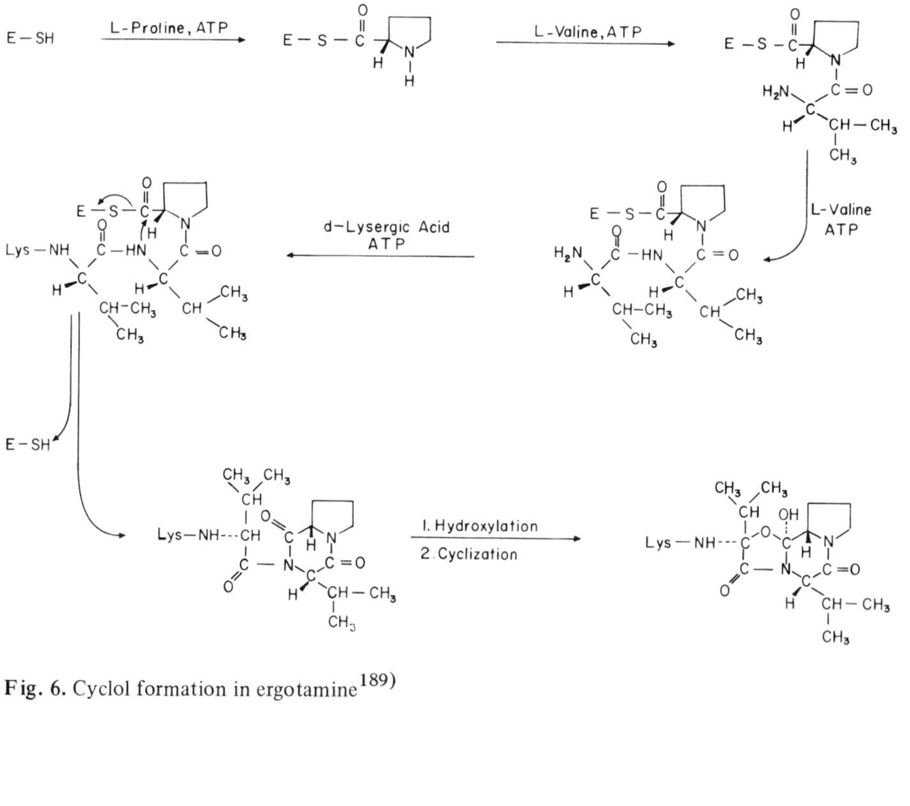

Fig. 6. Cyclol formation in ergotamine[189]

to lysergic acid (XV), paspalic acid occurs freely in the medium at higher concentrations.

Lysergic acid (XV) is a key substrate for the chemical synthesis of ergot alkaloids and their derivatives. At present it is formed both by isomerization of paspalic acid (XIV) and also by hydrolysis of the α-hydroxyethylamide of lysergic acid (XVIII). Both the initial substances are prepared by fungal cultivation. The natural derivatives of lysergic acid are the amides. Their formation requires activation of the carboxyl group, most probably as the coenzyme-A thiolester. The cell converts free lysergic acid to the amide[4]. The amide of lysergic acid is also formed by secondary chemical degradation of the hydroxyethylamide of lysergic acid (XVIII)[4, 75, 76].

Almost 20 peptides of lysergic acid are known; none the less, the biosynthesis of these ergopeptines remains obscure. Their formation is far more sensitive to cultivation conditions than the formation of clavines. The peptide part of 8-ergolene contains two structurally unusual components, α-hydroxy-α-amino acid and cyclol[12, 13, 77]. In the formation of cyclol (Fig. 6) protons pass from NH-group to the carboxyl. Simultaneously, a new N–C bond is formed, along with a new asymetric center with a tertiary OH group. The hydroxyl of the α-hydroxyalanine and α-hydroxyvaline parts of the peptide alkaloids is apparently incorporated into peptide chain, as it is improbable that the organism would synthesize three amino acids which do not occur in nature, i.e., L-α-hydroxyvaline, L-α-hydroxyalanine, and L-α-hydroxy-α-aminobutyric acid[78].

The discovery that L-proline-U-^{14}C is the precursor of the proline part of the ergotoxine peptide[79] and that prolyldiketopiperazines are incorporated into peptide alkaloids contributed to partial clarification of the mechanism of the biosynthesis of peptide alkaloids. Maier et al.[80] demonstrated that the formation of the peptide chain of ergoline begins with proline. Phenylalanine and proline are the most effective precursors of the peptide part of ergotamine[82]. Peptides are not precursors of the peptide part of ergotoxines. Submerged cultures of *Claviceps purpurea* producing ergocornine and ergocryptine split added peptides to the amino acids before incorporation into the alkaloids[83]. Condensation of the precursors of peptide alkaloids occurs through the activated intermediates. Extracts of *Claviceps* catalyze, e.g., the formation of the CoA-thiolester of lysergic acid[84]. It is, however, possible to assume the existence of other acylthiokinases activating further substrates. Inhibitors of proteosynthesis, such as cycloheximide, do not affect the formation of the peptide part of ergotoxine alkaloids[35]. The synthesis of the peptide part of ergoline is in all probability a nonribosomal process. It does not, however, proceed arbitrarily, in contrast to some peptides[10], which are formed by spontaneous condensation of amino acids, as, similarly as with peptide antibiotics, this process occurs on a multienzyme complex[85]. The tripeptide is not a free intermediate of this biosynthesis.

If we evaluate ergot alkaloids as secondary content substances, then their formation requires the participation of specific synthetases[87]. The number of isolated enzymes which catalyze the synthesis of ergot alkaloids is very small; however, two of the isolated enzymes fulfilled theoretical predictions: dimethylallyltryptophan synthetase[59, 86, 88] and chanoclavine cyclase[86, 89]. Dimethylallyltryptophan synthetase (dimethylallylpyrophosphate: tryptophandimethylallyl transferase) catalyzes the first specific reaction of the formation of 8-ergolene and is responsible for the introduction of the isoprene residue into position C(4) of tryptophan. Analogues of tryptophan increase the activity of this enzyme[90, 91]. The dimethylallyltryptophan (4-isoprenyltryptophan) produced is 5-10 times more effective as a precursor of clavine alkaloids than tryptophan[92]. Robbers and Floss[93] demonstrated accumulation of dimethylallyltryptophan under conditions of anaerobic cultivation of the fungus *Claviceps*. A further enzyme, chanoclavine-I-cyclase, catalyzes the cyclization of chanoclavine-I to the four-ring 8-ergolenes, i.e., to elymoclavine and agroclavine, both individually[94] and simultaneously[95].

This organisms producing clavine alkaloids are biocybernetically similar. They contain similar genes; the manner of their deactivation and activation in dependence on changes in medium and cell structures is also similar. In contrast, the biocybernetics of the organisms producing peptide ergot alkaloids are much more specific. This is also reflected in the fact that peptide alkaloids have so far been found only for one species, i.e., *Claviceps purpurea.*

Fungi and higher plants form ergolenes from the same precursors. In this connection it is important to determine whether the ergolene pathway was formed twice in nature independently or only once, the necessary genetic information being transferred from the fungus to the plant or vice versa. The concept of universal existence and limited expression of this genetic information is less probable.

5.2 Tryptophan

Tryptophan is the central precursor of ergot alkaloids[96]. Teuscher[47, 97] demonstrated active transfer of tryptophan from the medium to the mycelium of the *Claviceps* cultures and the dependence between this transport and the alkaloid formation. Mycelium growing in the initial phase of its development in excess tryptophan exhibits active alkaloid synthesis in a later phase of the process. Possible differences in nutrient acceptance by young and old cells could also be mentioned here. However, much data obtained in fungi studies indicate that the differences in assimilation and metabolism of amino acids by cells of the mycelium of various ages are negligible[27, 98]. Excessive accumulation of free tryptophan in mycelium affects the cell regulation systems, e.g., through induction of various enzymes in the synthesis of secondary content substances[99]. The absence of induced enzyme activity in a rapidly growing culture is attributed to catabolite repression[100-102].

Teuscher[97] found intense metabolism of tryptophan in the phase of intense alkaloid synthesis for a *Claviceps purpurea* culture. The formation of tryptophan in ergot strains occurs in the usual manner through chorismic and anthranilic acids[103]. The absence of Zn^{2+} and K^{+} decreases the synthesis of tryptophan. The participation of Zn^{2+} in increasing the level of cyclic AMP through inhibition of phosphodiesterase or activation of adenylate cyclase is also worth noting[104]. *Claviceps* strains degrade tryptophan in various ways, both common and unusual[47]. In intense formation of alkaloids, degradation of tryptophan through kynurenine does not play an important role[47, 105]. Under these conditions tryptophan is employed primarily for synthetic processes.

In recent years, enzymes used in the biosynthesis of tryptophan have been a subject of interest, especially their genetics and regulation possibilities. With most of the systems studied so far, the formation of enzymes of tryptophan metabolism is controlled by repression. Work on enzyme aggregation[106, 107] contributed to clarification of the basis of enzyme regulation. Various types of enzyme aggregate have been described, e.g., for *Neurospora* and other fungi[107, 108]. In each aggregate the activity of one or more enzymes depends on enzyme aggregation. In study of the relationship between genes and enzymes it was found that point mutation can be employed to attein simultaneously the loss of activity of several enzymes as a result of interference between the product of the gene and the aggregation.

The role of tryptophan in the synthesis of ergot alkaloids is not unambiguous. For several types of *Claviceps* it was not possible to find a relationship between the content of tryptophan in the mycelium and the alkaloid formation[109, 110] or increased biosynthetic activity after addition of tryptophan[111]. Gröger and Erge[112] attribute this to the rapid change of tryptophan through other metabolic pathways. If tryptophan is not a component of the nutritive medium, the cell forms it endogenously, with feedback control[113, 114]. If the mycelium grows on a medium containing tryptophan and is transferred after cessation of growth to a medium which does not contain tryptophan, alkaloid formation is more intense. The level of tryptophan in the mycelium grown in the presence of tryptophan is 2–3 times higher than in the mycelium from media without tryptophan.

The content of free and protein tryptophan is higher for cultures of *Claviceps purpurea* than for cultures of *Claviceps paspali*[36]. The ratio of the concentration of tryptophan in the cell pool to the concentration of protein tryptophan of the mycelium corresponds to a value of 1.8 for alkaloid production and to a value of 0.5 for nonproductive cultures. The decrease in the content of protein tryptophan in the mycelium has a favorable effect on the biosynthesis of alkaloids[115]. According to Erge et al.[35] the protein tryptophan is used for alkaloid formation. Radioisotope experiments, studying the fate of tryptophan added and endogeneously synthesized, indicated the existence of only one tryptophan pool in the cell[116]. The level of free tryptophan in the cell attains a maximum just before the beginning of alkaloid synthesis[65]. Řeháček and Malik[115] described three characteristic phases of the endogeneously formed tryptophan cell pool: the decrease phase, the increase phase and the excess phase. The boundary between the decrease phase and the increase phase is a suitable region for study of continuous alkaloid production.

The utilizability of endogeneous tryptophan is indicated by the levels of tryptophan synthetase and tryptophan oxygenase. Induction of tryptophan synthetase precedes induction of tryptophan oxygenase[33, 36]. Tryptophan synthetase is derepressed by imidazolylglycerol phosphate[117]. The decrease in the concentration of tryptophan in the cell pool is accompanied by increased activity of tryptophan synthetase. The activity of tryptophan oxygenase is higher in the presence of higher concentrations of free mycelium tryptophan[36], which probably protects tryptophan oxygenase against denaturization and proteolysis[118, 119].

Tryptophan is not only the direct precursor of ergot alkaloids, but is also a factor in the induction and depression of enzymes synthesizing alkaloids[10, 91, 120]. Vining[66] feels that tryptophan stimulates alkaloid formation especially through increased activity of the alkaloid enzyme system. Exogeneous tryptophan removes inhibition of alkaloid production by phosphate consisting of depressed tryptophan synthesis[65] and of activation of the enzymes splitting alkaloids[52]. Inorganic phosphate, which in greater concentrations completely inhibits alkaloid formation[33, 65, 121, 122], represses dimethylallyltryptophan synthetase[91] and chanoclavine cyclase[89]. Added tryptophan or thiotryptophan derepresses these enzymes. In its stimulating effect on alkaloid synthesis, tryptophan can be replaced by its nonmetabolized structural analogues[123]. On the other hand, the relationship of alkaloid formation to the metabolism of endogeneous tryptophan[36, 115, 124] stimulates consideration of the role of alkaloid biosynthesis in the regulation of the level of endogeneous tryptophan.

5.3 Gibbs Free Energy

The Gibbs free energy balance in the formation of ergot alkaloids[50] indicates that one of the key reactions in the biogenesis of ergot alkaloids is the formation of dimethylallylpyrophosphate (Fig. 7). This energetically demanding reaction precedes enzyme allylation of tryptophan in position C(4) which is poorly accessible chemically. In contrast, the subsequent cycle of clavine alkaloids is part of the endogenic metabolism releasing the energy necessary for maintaining the integrity and viability of the cells under conditions of limited growth or even under conditions in the absence of growth and

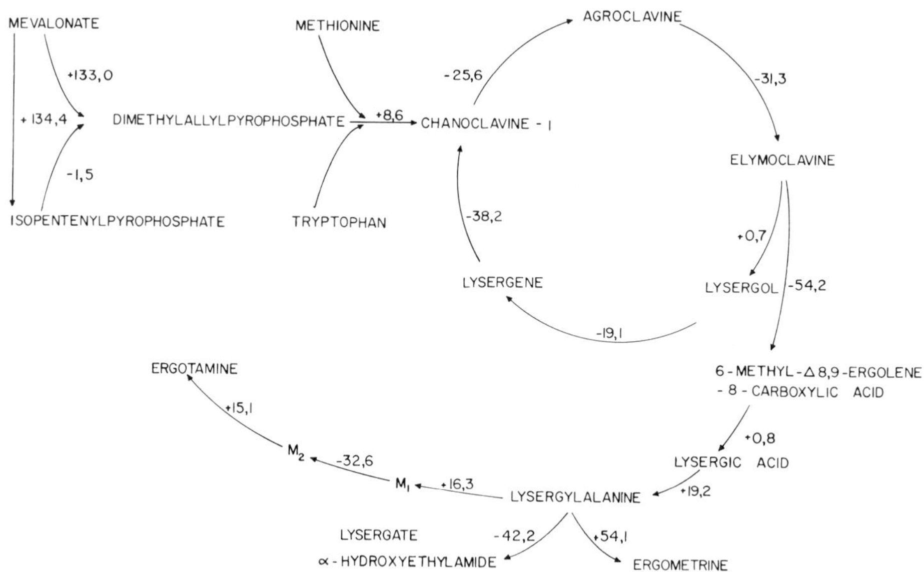

Fig. 7. Mutual relationships and balance of the Gibbs free energy (free enthalpy) in the ergot alkaloid formation[50]. The numbers refer to the free energy charge $\Delta G^{\circ\prime}$ in kcal per mole reactants at 24 °C, pH 7.0

death of the mycelium. The energy of the clavine cycle is reminiscent of the "energy of maintenance"[125] taking part in the resynthesis of the macromolecular components of the cell. This assumption is supported by the free energy balance of the formation of clavine alkaloids, by the low activity of the citrate cycle during the synthesis of clavines[36] and also by data indicating that the energy required for the turnover of the macromolecules is derived primarily from sources which the cell does not drain for its growth[57]. Considering that the evolution pressure results in types of cells marked by especially economical use of nutrients, it is possible to consider the positive role of the clavine cycle in the evolution of producers of clavine alkaloids. The manner of transformation of the biological energy of the clavine cycle is common for various taxonomically different organisms[50]. The formation of clavine alkaloids is accompanied by a decrease in the free energy with a tendency to achieve higher entropy. The molecules of derivatives of lysergic acid represent a more regular arrangement than clavine molecules. In the formation of derivatives of lysergic acid the free energy increases and entropy decreases.

The free energy balance in the formation of ergot alkaloids contributes, among other things, to the explanation of the frequent occurrence of clavines and of the α-hydroxyethylamide of lysergic acid and of the relatively rare occurrence of ergometrine and ergotamine in ergot cultures. An important role in the occurrence of clavine alkaloids in nature is also played by other factors, e.g., the mechanisms of *Claviceps paspali,* which protects sclerotium from deterioration through soil microorganisms, insects and root excreta of plants[131]. *Claviceps purpurea* lacks these protective mechanisms in its sclerotium; its sclerotium survives in the soil for approximately six months.

5.4 Regulation

Biosynthesis of alkaloids is regulated genetically as part of programs for culture differentiation and development. The assumed genetic connection of structural genes of the alkaloid system in a regulating unit of the operon contributes to clarification of the regulating mechanisms. Alkaloid synthesis and syntheses of some primary metabolites are controlled by common mechanisms. In the beginning of the alkaloid formation, the production of some primary metabolites is either blocked or inhibited. The sensitivity of some enzymes to activity regulation decreases during cell ageing. A condition for regulation of the biosynthesis of ergot alkaloids is knowledge of the physiology of the formation of these substances.

All cells contain more genes than enzymes. Regulating moments are decisive for gene manifestation. For example, a suitable physiological state of the cell is necessary for initiation of the action of genes controlling alkaloid formation. Genes involved in alkaloid synthesis are under rigorous metabolic control[136]. Many stages of alkaloid formation are not specific only for this biosynthetic pathway[127]. Some reactions of alkaloid synthesis are catalyzed by enzymes that also take part in other stages of the cell metabolism. In addition, reactions in which enzymes do not participate must not be neglected.

Alkaloids are not produced in cells with a lack of precursors or in which the alkaloid synthesis enzymes are repressed. The formation of alkaloid enzymes and thus also the production of alkaloids involve a derepression or induction process. Cells that do not form alkaloids can form partial sections of the alkaloid pathway. Then it would be possible to supplement the remaining sections in the sequence required by a mutation of one gene and thus complete the alkaloid synthesis. Mutation of one gene can also lead to the formation of new compounds and to structural modifications inside the cell. For example, cell components that were segregated in the parent cell can interact in the mutant. This general opinion has been supported by some observations[96)]. The feedback in the regulation of the ergot alkaloid synthesis results from the effect of the accumulated alkaloids. Dimethylallyltryptophan synthetase, the first enzyme of the alkaloid synthesis, is inhibited by agroclavine and elymoclavine, i.e., by the final products of the pathway of clavine alkaloid formation[59, 128)].

An indispensable part of the study of ergot alkaloids is consideration of the mechanism controlling the biosynthesis of tryptophan. Tryptophan represses all enzymes of its metabolic pathway and can also repress the enzyme-forming system of participating enzymes. The structural genes of the tryptophan system are genetically coupled and form a regulating unit[129)]. In wild strains, two regulating mechanisms have been found: inhibition of anthranilate synthetase by the feedback and enzyme repression by tryptophan. In mutants requiring tryptophan, the synthesis of enzymes is derepressed under conditions with limited tryptophan access; on blocking of the tryptophan synthesis more of the enzymes are liberated and repression then occurs due to their excess. Tryptophan has no effect on the feedback inhibition at the anthranilic acid site in some mutants[130)]. By mutation of the tryptophan operator, a mutant with constitutive enzyme synthesis can be obtained[132)]. In such a mutant either the gene has lost its ability to accept the repressor or the cell forms a defective tryptophan enzyme repressor. The inductor remains free and the enzyme biosynthesis proceeds constitutively and independently of the tryptophan requirement. In yeast, the tryptophan synthesis is regulated primarily by inhibition of anthranilate synthetase by the tryptophan feedback. Enzyme repression has no effect, or only a negligible effect[129)]. Anthranilic acid and indole act as precursors and interfere in the feedback inhibition. Higher concentrations of anthranilic acid have an inhibiting effect. Therefore, mutants accepting a higher level of anthranilic acid have been prepared. A combination of anthranilic acid with indole was more suitable than separate addition of these precursors[134)].

The increasing tryptophan level in the cell pool of a submerged *Claviceps* culture during the exponential development phase documents looser control of the tryptophan synthesis[36, 114, 135, 136)]. The biosynthetic stage at the boundary between the phases of the decrease and increase in the tryptophan level in the cell pool is a promising model for study of the regulation of the ergot alkaloid formation and for monitoring of continuous production of these metabolites[124)].

In following the physiology of the formation of ergot alkaloids, the metabolic pathways of histidine and tryptophan are worth studying, as they involve not only repression and feedback inhibition[129, 130, 137–139)], but also cross pathway regulation[36, 140)]. The alkaloid yield can be increased by adding amitrol (3-amino-1,2,4-triazol) to a three-day old submerged culture of *Claviceps purpurea.* Amitrol competitively inhibits the enzyme of the histidine pathway, imidazolylglycerolphosphate dehydratase[141)]. The effect of amitrol is manifest in accumulation of imidazolylglycerol in the cultivation medium, in a change in the protein and free tryptophan levels in the mycelium, in an increase in the alkaloid production and in a change in the alkaloid spectrum[140]. The change in the alkaloid spectrum caused by amitrol and the fact that the parasitic sclerotia of *Claviceps purpurea* contain alkaloids from the ergotoxine series, whereas the corresponding saprophytic strain forms lysergic acid and chanoclavine[142] support the opinion that the ergot alkaloids do not belong among decisive taxonomic parameters[140, 143]. On the other hand the cell ricinoleic acid is not only a chemotaxonomic parameter of the production culture of *Claviceps purpurea*[144] and an indicator of sclerotial type growth[145], but also an indicator of the ability of the strain to produce alkaloids[146].

Oxygenases play a certain role in the formation of alkaloids[147]. The favorable effect of aeration and the culture demand for molecular oxygen during the alkaloid synthesis can be attributed to these enzymes. Oxygenases contribute to the effective nourishing of the mycelium by the tryptophan precursor and to the effect of tryptophan as an inductor of certain enzymes. If oxygenases are inhibited, the cell has more tryptophan available for repression of the enzymes and the alkaloid biosynthesis is suppressed.

Ergot alkaloids participate in regulation of the formation of the enzyme-forming system[33, 148]. This role of alkaloids can be considered in terms of the induction hypothesis (Jacob and Monod), according to which the constitutive state predominates and requires an inductor binding the repressor before the enzyme-forming system begins to operate. The constitutive strain forms the inductor endogenously and the enzyme is formed automatically, provided that suitable cultivation conditions are created. *Claviceps paspali* does not form the inductor endogenously and the latter must be added[24, 112]. Tryptophan can be considered as an inductor[10, 120]. The tryptophan added is bound to the repressor and thus enables enzyme formation. However, tryptophan can also affect the alkaloid formation as a stimulation of the activity of the enzyme-forming system. Tryptophan and 5-methyltryptophan are effective initiators of dimethylallyltryptophan synthetase[91]. Thiotryptophan stimulates alkaloid formation more effectively than tryptophan[149].

The carbohydrate metabolism proceeds almost identically in productive and nonproductive *Claviceps* cultures up to the oxidative decarboxylation of pyruvate to acetylcoenzyme A. The most effective component in the carbohydrate degradation is the tricarboxylic acid cycle. The activity of its key enzymes is one to two orders of magnitude higher than that of the glycolytic and hexosomonophosphate enzymes. With citrate synthetase, a negligible activity has been found in a productive *Claviceps* culture and 3- to 5-times higher activity in a strain with a very low alkaloid content. Accumulation of citrate suppresses the citrate synthetase activity and activates acetylcoenzyme A carboxylase[150]. Under these conditions a greater amount of malonylcoenzyme A is formed and a greater number of condensation reactions takes place. During the alkaloid synthesis acetylcoenzyme A enters the tricarboxylic acid cycle to a limited degree and takes an alternative path to the synthesis of fatty acids[151]. The lipid content in the mycelium increases almost parallel to the activity of acetylcoenzyme A carboxylase and the total alkaloid level[33, 43].

In a submerged culture of *Claviceps purpurea* with high alkaloid yields, acetylcoenzyme A carboxylase was cytochemically localized in the endoplasmic reticulum of sclerotial cells[51]. The activity of this enzyme is structurally bound in membranous fibers of the endoplasmic reticulum which later participate in the formation of vacuoles. Acetylcoenzyme A carboxylase has not been found in the endoplasmic reticulum of nonvacuolized fibrous hyphae, ovoid spores with numerous lipidic droplets and cells of a nonproductive strain.

In the beginning of submerged cultivation of the *Claviceps* strain with intense cell proliferation, the tricarboxylic acid cycle is the predominant amphibolic pathway. The culture gains energy and intermediates for synthetic processes in the subsequent alkaloid phase chiefly through the glyoxylate cycle[152, 192]. During the alkaloid phase, the glyoxylate cycle simultaneously enables the introduction of pyruvate into cells. 2-Ketoglutaric, isocitric, citric, and fumaric acids stimulate the synthesis of alkaloids in cells with an intense glyoxylate cycle and decreasing activity of isocitrate dehydrogenase[152]. The glyoxylate cycle contains some reactions in common with the tricarboxylic acid cycle. In eucaryon cells the two cycles are physically separated. Most enzymes of the glyoxylate cycle are located in organelles termed glyoxysoms[153–155]. Mitochondria are the components of the enzymes of the tricarboxylic acid cycle. Possible participation of glyoxysomes and mitochondria in the regulation of alkaloid biosynthesis cannot be excluded.

Chemical energy is transferred in all enzymatic reactions. Regulating mechanisms based on a general form of stored energy also participate in the control of alkaloid synthesis. Atkinson[156, 157] characterizes this type of control as the activation and inhibition of the enzymes of the primary metabolism by relative levels of ATP, ADP, and AMP in the cell. The energy content is measured quantitatively as the ratio [ATP] + 1/2 [ADP]/[ATP] + [ADP] + [AMP]. A high energy content inhibits some enzymes of the primary metabolism. The biosynthesis of ergot alkaloids is inhibited by inorganic phosphate at concentrations that are optimal for the culture growth or do not inhibit the growth[158]. A high phosphate level in the medium activates the enzymes degrading alkaloids[52] and stimulates the formation of ATP and an increase in the energy content. An increase in the ATP level in the cell pool is accompanied by inhibition of citrate synthetase, phosphoenolpyruvate carboxylase and phosphoenolpyruvate carboxykinase and participates in the suppression of the activity of the tricarboxylic acid cycle[159]. A decrease in the activity of the tricarboxylic acid cycle is favorable for the alkaloid formation. Adenosinetriphosphatase in cell-free preparations of *Claviceps paspali* is inhibited by added ergometrine[33, 34]. This finding is the first illustration of the potential role of ergot alkaloids in the producent energy metabolism. The finding that an increasing concentration of ATP in the mycelium pool and decreased utilization of ATP favorably affect the alkaloid formation[34] suggests that the production of energy is not a limiting factor in the alkaloid biosynthesis. As can be expected, a low concentration of AMP supports the use of acetylcoenzyme A for biosynthesis of fatty acids and alkaloids, at the expense of the tricarboxylic acid cycle inhibited by a high ATP concentration. Syntrophic effects must also be taken into consideration. Fungi, in contrast to bacteria, have a greater amount of substances available, from which they can energetically supply, e.g., 72-h cultivation.

Temporary changes in the biosynthesis of ergot alkaloids can be induced by manipulation of cultivation conditions. Kelleher et al.[160] increased the production of alkaloids twice or three-times by adding 0.075% Tween 80 to the cultivation medium containing a submerged culture of *Claviceps paspali.* However, Tween 80 at this concentration causes excessive foaming of the cultivation medium and the suppression of foaming in turn inhibits the alkaloid synthesis. Dimethylsulfoxide (1%) increases the alkaloid yield by 50% without foaming of the cultivation liquid. Both these surfactans affect the permeability of the cytoplasmic membrane. By disturbing the permeation barrier, the cells increase the liberation of alkaloids from the region of their synthesis and thus prevent possible feedback inhibition. Surfactans do not affect the biosynthetic pathway[161]. The shortening of the alkaloid formation phase by the addition of Tween 80 is accompanied by a shift in the organic acid and amino acid level in the cell pool[162]. From this finding it follows that the alkaloid biosynthesis is regulated by the level of the intracellular precursors of the primary metabolites.

An increased concentration of numerous metabolites in the cell pool can also be attained by increasing the medium osmolarity[163]. An increased concentration of free metabolites in the cell prevents their further synthesis (feedback control). Membranous systems of eucaryons can maintain a relatively high level of native proteins in the cell, but cannot maintain a high level of low-molecular metabolites. Development pressure favors low concentrations of metabolites. Ergot alkaloids are produced only in media with a high osmotic pressure[47,48]. Puc a. Sočič[164] obtained a maximum alkaloid yield with ergocornine strain of *Claviceps purpurea* on a medium containing 20% mannitol or 10% mannitol and 2% sodium chloride, i.e., with a final osmolarity of the cultivation medium of about 1 osmole (kg $H_2O)^{-1}$.

Data on the effect of aeration of the cultivation liquid on the alkaloid production are controversial. A decreased respiration is given as one of the possible ways of increasing the alkaloid formation[165]. However, the opposite opinion has also been expressed[159, 166].

Permanent changes favoring alkaloid synthesis can be induced by manipulating the strain genetic structure[167, 168]. In preparation of productive mutants of *Claviceps,* doses of a mutagen leading to a percentage of surviving cells lower than 0.5 gave good results. Mutants obtained by irradiation with UV light were more stable than those prepared by treatment with ethylmethansulphonate. It is more difficult to affect the alkaloid spectrum by mutagenic treatment than to increase the alkaloid production.

5.5 Strain Stability

Productive strains often lose the ability to form alkaloids after a series transfer of the spores or of vegetative mycelium. The loss of production ability is sometimes accompanied by changes in the culture morphology[30]. The fact that strain cells are polynuclear contributes to the explanation of the phenomenon of strain degeneration. The physiological manifestations of polynuclear cells result from combined genetic cooperation of various nuclei in the common cytoplasmic field. During the cell growth the individual nuclei may divide at various rates. A shift in the genetic configuration thus occurs, leading to a shift in the phenotypical manifestation of the population. Jinks[169] classified the differences in the intensity of the alkaloid formations in the category of cytoplasmic variations. Changes in the strain ability to produce alkaloids occur not only due to genetic changes, but also due to exceptional sensitivity of the strain to changes in the cultivation medium. Living cells are an oscillating system that constantly passes through series of states and is simultaneously an active manipulator of the environment.

6 Physiological Role

Biosynthesis of ergot alkaloids plays a role in the production organism physiology by maintaining certain mechanisms of the metabolism and cell division in an operative state, especially under conditions of the limited growth of the organism. This concept is supported by experimental results that confirm a considerable protein turnover during the phase of alkaloid formation[33, 170], by the finding that nongrowing cells die if the metabolism of the secondary content substances is inhibited in them[87] and finally by the proof of inverse dependence of the production of secondary content substances on the cell differentiation[171]. On cessation of the process of cell division, the alkaloid synthesis is directly related to the viability of the culture. Highly productive mutants of *Claviceps* strains are an exception to this consideration; they are marked by a lower viability and occur only exceptionally under natural conditions.

Ergot alkaloids are not waste, physiologically inert products of cell metabolism[33, 34, 124]. Mothes[172] considers their formation as a detoxificating process. Alkaloid synthesis is one of the ways of removal of accumulated metabolic intermediates. This process enables the reactions in which these intermediates are formed to remain in an operative state and thus participates in the regulation of the primary metabolism.

In a productive cell, ergot alkaloids increase the activity of tryptophan synthetase and acetylcoenzyme A carboxylase and inhibit anthranilate synthetase, isocitrate lyase, malate synthetase and adenosine triphosphatase[34, 124, 136]. Suppression of the peptidase activity was also observed (Řeháček and Sajdl, unpublished). Ergometrine (10^{-4} M) added to a submerged culture of *Claviceps paspali* increases the alkaloid formation by 120%[34]. Ergotamine (5 x 10^{-4} M) and ergometrine (5 x 10^{-4} M) reduce the intensity of conidia formation in a productive culture of *Claviceps purpurea*[31].

The effect of tryptophan on the enzyme system in the ergot alkaloid synthesis[10, 41, 66] suggests participation of the alkaloids in the feedback inhibition and repression of the enzyme-forming system. At low concentrations of alkaloids and tryptophan, the reactions are ordered linearly or sequentially. The product of one reaction in the series is the substrate of the subsequent reaction[66]. The substrate (tryptophan) and the final product (alkaloid) do not compete for the site on the enzyme surface because their chemical structures are different and hence possible inhibition is allosteric and noncompetitive. On blocking of the alkaloid synthetase the enzyme-forming system is derepressed and the cell forms more enzymes. After formation of a higher enzyme level, the enzyme are repressed again.

Clavine alkaloids probably form a reservoir from which, in the course of evolution, interactions developed leading to the formation of peptide alkaloids. Alkaloids are not formed to perform a certain function, but under certain conditions some alkaloids "find" their function. This situation has an analogue in morphology. The formation of clavine alkaloids is a part of endogenous metabolism releasing the energy required for maintenance of the integrity and viability of cells under conditions of limited growth[140]. As part of important regions in the productive cell it is more resistant to regulation noise.

The occurrence of ergot alkaloids depends not only on genetic information, but also on the ability of the cell to tolerate these substances. Alkaloids confirm that metabolic development must be supplemented by coevolution of the tolerance mechanism toward some secondary content substances[173]. It can be generally stated that mechanisms developed in productive organisms against their own products are not different from those developing in other originally sensitive organisms.

7 Conclusion

Biosynthesis of ergot alkaloids is a genetically controlled enzymatic process. There is little experimental data on the type of genetic control, but they permit the assumption of the existence of a polygenic system as a universal model for the control of alkaloids and alkaloid phenotypes. Alkaloid formation thus lies at the level of the genetic information apparatus and its realization continues at the level of physiological realization.

It is justly assumed that the principal findings about the physiology of the alkaloid formation will be provided by the results of the study of the membrane processes in the productive organisms. Valuable data are also expected from enzyme cytology, chiefly because this discipline combines dynamic biochemistry and descriptive cytology in order to understand the cell from the point of view of functional properties and spatial organization of its components. Numerous metabolic functions are localized in cell structures and considerations on structures thus should become a part of functional analysis. This field includes the problem of compartmentation of enzymes and macromolecules, the problem of distribution of alkaloids and their transport at the molecular level. The fact that various kinds solve the same functional problems differently emphasizes the need for application of comparative functional anatomy of subcellular organization. Vacuoles – a polyvalent class of cell organelles – are also worth studying, as they participate in the degradation of cell components and are storage places for macromolecules. The first goal should be identification of the mechanism by which cytoplasm components are incorporated into the vacuole.

The complex relationship between primary metabolism and alkaloid formation has not yet been studied sufficiently extensively. Here attention could be paid especially to the following questions: Which factors are responsible for accumulation of a primary metabolite controlling the alkaloid synthesis? Is catabolite repression of alkaloid synthesis related preferentially to the intensity of the culture growth or to the specific source of carbon or nitrogen? What are the kinetic constants of alkaloid formation control? Are there some aspects of the primary metabolism and of the alkaloid synthesis that are mutually exclusive? If so, can they be made compatible by changing the control system?

Industrial process is generally developed on the foundations provided by basic research and empirical experience[174]. A condition for modern production of ergot alkaloids is knowledge of physiological controls, for rationalization of the process conditions including shortening of the initial nonproductive phase and for regulated preparation of highly productive mutants. Among the alkaloids, ergolene carboxylic acids (paspalic, lysergic) are in the center of attention at present. They are substrates for chemical synthesis of the classical ergot alkaloids and their derivatives[175]. Interest is growing in clavine alkaloids[15, 176, 177], especially in elymoclavine[178]. A specific point in this research is the preparation of ergot alkaloids by coreaction using a mixed culture of productive organisms[179] or of a productive organism and its natural competitive contaminant[180].

The newer ergot derivatives have opened up a whole new area of investigation into the pharmacologic activity for the ergots especially as it applies to the neuroendocrine system. There are several potential therapeutic uses for which the newer ergot alkaloids are currently being considered, including the treatment of parkinsonism, acromegaly, amenorrhea/galactorrhea syndrome, in the suppression of postpartum lactation, and possibly as an adjunct in the treatment of breast cancer and prostatic carcinoma.

The search for new natural alkaloids involves the testing of nontraditional organisms. The more related the organisms are, the greater is the probability of their forming the same alkaloids. The series of ergot alkaloids is widened by their semisynthetic analogues. Some of them are important practically or find use in the study of the relationships between the chemical structure and the biological effect.

In spite of past successes of ergot alkaloids in practice, the future of these pharmaceuticals depends on the results of basic research. With new discoveries, new unexpected problems will, of course, arise. For this purpose it is necessary to provide a certain material and mental capacity.

8 References

1. Berde, B., Schild, H.O. (eds.): Ergot alkaloids and related compounds. Berlin: Springer-Verlag 1978
2. Refai-El: Jpn. J. Microbiol. *14*, 91 (1970)
3. Hofmann, A.: Die Mutterkornalkaloide. Stuttgart: Enke 1964
4. Agurell, S.: Acta Pharm. Suec. *3*, 23 (1966)
5. Perlman, D.: ASM News *43*, 82 (1977)
6. Bové, F.J.: The story of ergot. Basel: S. Karger 1970
7. Stoll, A., Hofmann, A.: The alkaloids, Vol. VIII, p. 725. New York: Academic Press 1965
8. Stadler, P.A., Stütz, P.: The alkaloids, Vol. XV, p. 1. New York: Academic Press 1975

9. Woodward, R.B.: Angew. Chem. *68*, 13 (1956)
10. Bu' Lock, J.D., Barr, J.G.: Lloydia *31*, 342 (1968)
11. Pöhm, M.: Monats. Chem. *108*, 393 (1977)
12. Wrinch, D.: Nature *137*, 411 (1936)
13. Hofmann, A., Frey, A.J., Ott, H.: Experientia *17*, 206 (1961)
14. Floss, H.G.: Tetrahedron *32*, 873 (1976)
15. Floss, H.G.: J. Med. Chem. *17*, 300 (1974)
16. Schlientz, W. et al.: Helv. Chim. Acta *47*, 1921 (1964)
17. Weber, H.P., Petcher, T.J.: The shape of ergotamine and dihydroergotamine in relation to their interaction with α-adrenoceptors. In: Ergot alkaloids and related compounds. Berde, B., Schild, H.O. (eds.), p. 177. Berlin: Springer-Verlag 1978
18. Rothlin, E., Brügger, J.: Helv. physiol. pharmacol. Acta *3*, 519 (1945)
19. Rothlin, E.: Bull. schweiz. Akad. med. Wiss. *2*, 249 (1946/47)
20. Abe, M.: Ann. Rep. Takeda Res. Lab. *10*, 110 (1951)
21. Abbour-Chaar, C.I., Brady, L.R.: Lloydia *24*, 89 (1961)
22. Abe, M., Yamatodani, S.: Preparation of alkaloids by saprophytic culture of ergot fungi. In: Progress in industrial microbiology. Hochenhull, D.J.D. (ed.), Vol. 5, p. 205. London: Heywood 1964
23. Abe, M. et al.: J. Agr. Chem. Soc. Jpn. *41*, 68 (1967)
24. Taber, W.A.: Lloydia *30*, 39 (1967)
25. Mantle, P.G., Nisbet, L.J.: J. gen. Microbiol. *93*, 321 (1976)
26. Turian, G.: Experientia *32*, 989 (1976)
27. Zalokar, M.: Amer. J. Bot. *46*, 555 (1959)
28. Turian, G.: Trans. Brit. Mycol. Soc. *64*, 367 (1975)
29. Morton, J.: Sci. Progress Oxford *55*, 597 (1967)
30. Pažoutová, S., Pokorný, V., Řeháček, Z.: Can. J. Microbiol. *23*, 1182 (1977)
31. Řeháček, Z. et al.: Can. J. Microbiol. *20*, 1223 (1974)
32. Sekiguchi, J., Gaucher, G.M.: Appl. Environ. Microbiol. *33*, 147 (1977)
33. Řeháček, Z. et al.: Appl. Microbiol. *22*, 949 (1971)
34. Řeháček, Z. et al.: Fol. Microbiol. *17*, 308 (1972)
35. Erge, D., Wenzel, A., Gröger, D.: Biochem. Physiol. Pflanzen *163*, 288 (1972)
36. Řeháček, Z. et al.: Fol. Microbiol. *16*, 35 (1971)
37. Pöhm, M., Kolassa, H., Jentzsch, K.: Sci. Pharm. *44*, 32 (1976)
38. Kaplan, H. et al.: Lloydia *32*, 489 (1969)
39. Rothe, K., Fritsche, W.: Arch. Mikrobiol. *58*, 77 (1967)
40. Řeháček, Z. et al.: Can. J. Microbiol. *23*, 596 (1977)
41. Taber, W.A., Vining, L.C.: Can. J. Microbiol. *9*, 1 (1963)
42. Voříšek, J., Ludvík, J., Řeháček, Z.: J. Bacteriol. *120*, 1401 (1974)
43. El-Sayed, D.H., Řeháček, Z.: Zbl. Bact. Abt. II *133*, 551 (1978)
44. Whiteworth, D.A., Ratledge, C.: J. Gen. Microbiol. *88*, 275 (1975)
45. Drew, S.W., Demain, A.L.: Ann. Rev. Microbiol. *31*, 343 (1977)
46. Vining, L.C., Nair, P.M.: Can. J. Microbiol. *12*, 915 (1966)
47. Teuscher, E.: Beiträge zur Physiologie der Alkaloidbildung unter besonderer Berücksichtigung der Biogenese der Mutterkornalkaloide. Habilitationsschrift. Greifswald, Ernst-Moritz-Arndt-Universität 1965
48. Tonolo, A.: Ann. Ist. Super Sanita *3*, 613 (1967)
49. Mátlová, S., Řeháček, Z.: Fol. Microbiol. *21*, 211 (1976)
50. Řeháček, Z., Sajdl, P., Křemen, A.: Biotechnol. Bioeng. *15*, 207 (1973)
51. Voříšek, J., Řeháček, Z.: Arch. Microbiol. *117*, 297 (1978)
52. Robbers, J.E., Eggert, W.W., Floss, H.G.: Lloydia *41*, 120 (1978)
53. Birch, A.J., McLoughlin, B.J., Smith, H.: Tetrahedron Lett. *7*, 1 (1960)
54. Gröger, D. et al.: Z. Naturforsch. *15*b, 141 (1960)
55. Taylor, A.L., Ramstad, E.: Nature *188*, 494 (1960)

56. Tyler, V.E.: J. Pharm. Sci. *50*, 629 (1961)
57. Plieninger, H., Immel, H.: Lieb. Ann. Chem. *706*, 223 (1967)
58. Wenkert, E., Sliwa, H.: Biorg. Chem. *6*, 443 (1977)
59. Heinstein, P.F., Lee, S.L., Floss, H.G.: Biochem. Biophys. Res. Commun. *44*, 1244 (1971)
60. Lee, S.I., Floss, H.G., Heinstein, P.F.: Arch. Biochem. Biophys. *177*, 84 (1976)
61. Petroski, R.J., Kelleher, W.J.: FEBS Lett. *82*, 55 (1977)
62. Petroski, R.J., Kelleher, W.J.: Lloydia *41*, 332 (1978)
63. Anderson, J.A., Saini, M.S.: Tetrahedron Lett., p. 2107 (1974)
64. Floss, H.G.: 4th Intern. Symp. Biochem. Physiol. Alkaloide. Halle, Berlin: Academie-Verlag 1969
65. Robbers, J.E. et al.: Bacteriol. *112*, 791 (1972)
66. Vining, L.C.: Can. J. Microbiol. *16*, 473 (1970)
67. Lynen, F.: Biochem. J. *102*, 381 (1967)
68. Gupta, A.R., Rao, K.K.: Ind. J. Biochem. Biophys. *14*, 84 (1977)
69. Abe, M.: Abh. Dt. Akad. Wiss. Berlin, p. 393 (1966)
70. Wilson, B.J. et al.: Biochim. Biophys. Acta *252*, 348 (1971)
71. Beliveau, J., Ramstad, E.: Lloydia *29*, 234 (1966)
72. Jindra, A., Ramstad, E., Floss, H.G.: Lloydia *31*, 190 (1968)
73. Gröger, D., Erge, D., Floss, H.G.: Z. Naturforsch. *21* b, 827 (1966)
74. Floss, H.G.: Chem. Commun., p. 804 (1967)
75. Ramstad, E.: Lloydia *31*, 327 (1968)
76. Kleinerová, E., Kybal, J.: Fol. Microbiol. *18*, 390 (1973)
77. Shemyakin, M.M. et al.: Tetrahedron Lett., p. 701 (1962)
78. Cavender, F.L., Anderson, J.A.: Biochim. Biophys. Acta *208*, 345 (1970)
79. Gröger, D., Erge, D.: Z. Naturforsch. *25* b, 196 (1970)
80. Maier, W., Erge, D., Gröger, D.: Biochem. Physiol. Pflanzen *161*, 559 (1971)
81. Kybal, J., Kleinerová, E., Bulant, V.: Fol. Microbiol. *21*, 474 (1976)
82. Basset, R.A., Chain, E.B., Corbett, K.: Biochem. J. *134*, 1 (1973)
83. Gröger, D., Johne, S., Härtling, S.: Biochem. Physiol. Pflanzen *166*, 33 (1974)
84. Maier, W., Erge, D., Gröger, D.: Biochem. Physiol. Pflanzen *163*, 432 (1972)
85. Gröger, D.: Planta Medica *28*, 37 (1975)
86. Floss, H.G., Robbers, J.E., Heinstein, P.F.: Recent Adv. Phytoch. *8*, 141 (1974)
87. Weinberg, E.D.: Adv. Microbiol. Physiol. *4*, 1 (1970)
88. Maier, W., Gröger, D.: Biochem. Physiol. Pflanzen *170*, 9 (1976)
89. Erge, D., Maier, W., Gröger, D.: Biochem. Physiol. Pflanzen *164*, 234 (1973)
90. Schmauder, H.P., Gerullis, C., Gröger, D.: Biochem. Physiol. Pflanzen *170*, 201 (1976)
91. Krupinski, V.M., Robbers, J.E., Floss, H.G.: J. Bacteriol. *125*, 158 (1976)
92. Plieninger, H., Fischer, R., Liede, V.: Lieb. Ann. Chem. *672*, 223 (1964)
93. Robbers, J.E., Floss, H.G.: Arch. Biochem. Biophys. *126*, 967 (1968)
94. Gröger, D., Sajdl, P.: Pharmazie *27*, 188 (1972)
95. Sajdl, P., Řeháček, Z.: Fol. Microbiol. *20*, 365 (1975)
96. Mothes, K. et al.: Z. Naturforsch. *14* b, 49 (1959)
97. Teuscher, E.: Flora *155*, 80 (1964)
98. Bent, K.J.: Biochem. J. *92*, 280 (1964)
99. Bu'Lock, J.D., Powell, A.J.: Experientia *21*, 55 (1965)
100. Horowitz, N.H.: Biochem. Biophys. Res. Commun. *18*, 686 (1965)
101. Stadman, E.R.: Adv. Enzymol. *28*, 41 (1966)
102. Marshall, R. et al.: Arch. Biochem. Biophys. *123*, 317 (1968)
103. Gröger, D. et al.: Z. Naturforsch. *16* b, 432 (1961)
104. Chvapil, M.: Life Sci. *13*, 1041 (1973)
105. Matchett, W.H., DeMoss, J.A.: Biophys. *123*, 317 (1968)
106. DeMoss, J.A.: Biochem. Biophys. Res. Commun. *18*, 850 (1965)
107. Hütter, R., DeMoss, J.A.: J. Bacteriol. *94*, 1896 (1967)

108. DeMoss, J.A., Wegman, J.: Proc. Nat. Acad. Sci. U.S. *54*, 241 (1965)
109. Gjerstad, G.: J. Amer. Pharm. Ass. *48*, 443 (1959)
110. Gröger, D.: Pharmazie *15*, 715 (1960)
111. Gröger, D., Tyler, Jr., V.E.: Lloydia *26*, 174 (1963)
112. Gröger, D., Erge, D.: Pharmazie *19*, 775 (1964)
113. Wiley, W.R., Matchett, W.H.: J. Bacteriol. *95*, 959 (1968)
114. Lingens, F., Göbel, W., Üsseler, H.: Naturwiss. *54*, 141 (1967)
115. Řeháček, Z., Malik, K.A.: Fol. Microbiol. *16*, 359 (1971)
116. Hornemann, U. et al.: Arch. Biochem. Biophys. *131*, 430 (1969)
117. Turner, R., Matchett, W.H.: J. Bacteriol. *95*, 1608 (1968)
118. Fiegelson, P., Greengard, O.: J. Biol. Chem. *237*, 1908 (1962)
119. Knox, W.E., Mehler, A.H.: Science *113*, 237 (1951)
120. Floss, H.G., Mothes, U.: Arch. Mikrobiol. *48*, 213 (1964)
121. Mary, N.Y., Kelleher, W.J., Schwarting, A.E.: Lloydia *28*, 218 (1965)
122. deWaart, C., Taber, W.A.: Can. J. Microbiol. *6*, 675 (1960)
123. Robbers, J.E., Floss, H.G.: J. Pharm. Sci. *59*, 702 (1970)
124. Řeháček, Z. et al.: Ergot alkaloids formation in relation to the cell-pools of tryptophan and adenosine-5'-triphosphate. In: Genetics of industrial microorganisms. Actinomycetes and Fungi. Vaněk, Z., Hoštálek, Z., Cudlín, J. (eds.), p. 427. Amsterdam: Elsevier Publ. Comp. 1973
125. Gest, H.: Coupling of energy conversion and biosynthesis processes in growing bacteria. In: Genetics of industrial microorganisms. Bacteria. Vaněk, Z., Hoštálek, Z., Cudlín, J. (eds.), p. 131. Amsterdam: Elsevier Publ. Comp. 1973
126. Vining, L.C.: 1st Int. Symp. Genetics of Industrial Microorganisms. Abstr. Book, p. 144. Prague 1970
127. James, W.A.: Pharm. Pharmacol. *5*, 809 (1953)
128. Malik, V.S., Vining, L.C.: Can. J. Microbiol. *16*, 173 (1970)
129. Hütter, R.: Rep. Int. Atomic Energy Agency IAE/SM-134/31. Vienna 1971
130. Sommerville, R.L., Yanofsky, C.: J. Mol. Biol. *11*, 474 (1965)
131. Cunfer, B.M., Seckinger, A.: Mycologia *69*, 1142 (1977)
132. Miraga, S.: J. Med. Biol. *39*, 159 (1968)
133. Agurell, S.: Acta Pharm. Suecica *3*, 71 (1966)
134. Ebihara, Y., Niitsu, H., Terui, G.: J. Ferment. Technol. *47*, 733 (1969)
135. Lingens, F., Göbel, W., Üsseler, H.: Naturwiss. *54*, 141 (1967)
136. Schmauder, H.P., Gröger, D.: Biochem. Physiol. Pflanzen *169*, 201 (1976)
137. Cohen, J.F., Jacob, F.: Compt. Rend. Acad. Sci. Paris *248*, 3490 (1959)
138. Kovach, J.S. et al.: J. Bacteriol. *97*, 1283 (1969)
139. Morse, D.E., Yanofsky, C.: J. Mol. Biol. *44*, 185 (1969)
140. Řeháček, Z.: Physiology of formation of some antibiotics and ergot alkaloids. DSc Thesis, Prague, Inst. Microbiol. Czechosl. Acad. Sci. 1972
141. Hilton, J.: Modes of action of 3-amino-1,2,40triazole. In: Isotopes in weed research, p. 71. Vienna, Int. Atomic Energy Agency 1966
142. Mantle, P.G.: J. Reprod. Fertility *18*, 81 (1969)
143. Mothes, K.: Lloydia *29*, 156 (1966)
144. Weiblinger, K., Gröger, D.: Biochem. Physiol. Pflanzen *163*, 468 (1972)
145. Morris, L.J., Hall, S.W.: Trans. Br. Mycol. Soc. *53*, 441 (1969)
146. Řeháček, Z., Kozova, J.: Fol. Microbiol. *20*, 112 (1975)
147. Ogunlana, E.O., Ramstad, E., Tyler, V.E.: J. Pharm. Sci. *58*, 143 (1969)
148. Řeháček, Z., Malik, K.A.: Fol. Microbiol. *17*, 490 (1972)
149. Robbers, J.E., Floss, H.G.: Nova Acta Leopoldina 7, 243 (1974)
150. Atkinson, D., Goodwin, T.W.: In: Metabolic roles of citrate, p. 23. New York: Academic Press 1968
151. Che-Chang, H. et al.: Biochem. Biophys. Res. Commun. *28*, 682 (1967)
152. Řeháček, Z., Sajdl, P., Dutta, S.M.: Zbl. Bakt. Abt. II. (in press)

153. Müller, M., Hogg, J.F., DeDuve, C.: J. Biol. Chem. *243*, 5385 (1968)
154. Cooper, F.G., Beevers, H.: J. Biol. Chem. *244*, 5507 (1969)
155. Kobr, M.J., Vanderhaeghe, F., Combépine, G.: Biochem. Biophys. Res. Commun. *37*, 640 (1969)
156. Atkinson, D.E.: Ann. Rev. Microbiol. *23*, 47 (1969)
157. Atkinson, D.E.: Biochem. *7*, 4030 (1968)
158. Weinberg, E.D.: Dev. Ind. Microbiol. *15*, 70 (1974)
159. Atkinson, D.E.: Science *150*, 851 (1965)
160. Kelleher, W.J., Kim, B.K., Schwarting, A.E.: Lloydia *32*, 327 (1969)
161. Mizrahi, A., Miller, G.: Biochem. Bioeng. *10*, 102 (1968)
162. Řeháček, Z., Basappa, S.C.: Fol. Microbiol. *16*, 110 (1971)
163. Stebbing, N.: Bact. Rev. *38*, 1 (1974)
164. Puc, A., Sočič, H.: Eur. J. Appl. Microbiol. *4*, 283 (1977)
165. Windisch, S., Brown, W.: US. Pat. 2,936,266 (May 10, 1960)
166. Arcamone, F. et al.: Proc. Roy. Soc. London, Ser. B *155*, 26 (1961)
167. Kybal, J., Strnadová, K.: Čs. Farmacie *24*, 25 (1975)
168. Řeháček, Z. et al.: Pat. Appl. PV 2516-78 (1978)
169. Jinks, J.L.: Compt. Rend. Trav. Lab. Carlsberg, Ser. Physiol. *26*, 183 (1956)
170. Kobel, H., Schreier, E., Rutschmann, J.: Helv. Chim. Acta *47*, 1052 (1964)
171. Woodruff, H.B.: The physiology of antibiotic production. The role of producing organism. In: Biochemical studies of antimicrobial drugs. Newten, B.A., Reynolds, P.E. (eds.), p. 22. London: Cambridge University Press 1966
172. Mothes, K.: Nova Acta Leopoldina *7*, 403 (1976)
173. Hegnauer, R.: Nova Acta Leopoldina *7*, 45 (1976)
174. Udvardy, E.N. et al.: Conf. Med. Plants, p. 51. Marianské Lázně 1975
175. Schreier, E.: Helv. Chim. Acta *59*, 585 (1976)
176. Řeháček, Z. et al.: Pat. Appl. PV 2517-78 (1978)
177. Řeháček, Z., Sajdl, P.: Stud. Czech. Acad. Sci. (in press)
178. Choong, Ch. T., Sough, H.R.: Tetrahedron Lett. *19*, 1627 (1977)
179. Kobel, H.: Conf. Med. Plants, p. 72. Mar. Lázně 1975
180. Kaczmarek, F., Haniecka-Jaskulska, T.: Herba Polon. *20*, 264 (1974)
181. Abe, M.: 1st Int. Congr. Genetics of Industrial Microorganisms. Abstract Book, p. 149. Prague 1970
182. Basmadjian, G. et al.: Chem. Commun. 418 (1969)
183. Nelson, U., Agurell, S.: Acta Chim. Scand. *23*, 3393 (1969)
184. Pachlatke, P., Tobacik, C., Acklin, W., Arigoni, D.: Chimia *29*, 526 (1975)
185. Robinson, T.: The biochemistry of alkaloids, p. 77. Berlin: Springer-Verlag 1968
186. Voigt, R.: Pharmazie *23*, 353 (1968)
187. Weygand, F., Floss, H.G.: Angew. Chem. Intern. Ed. *2*, 243 (1963)
188. Jacobs, W.A., Gould, R.G.: J. Biol. Chem. *120*, 141 (1937)
189. Floss, H.G. et al.: Experientia *30*, 1369 (1974)
190. Turian, G.: Différenciation Fongique. Paris: Masson & Cie 1969
191. Smith, J.E., Berry, D.R.: An introduction to biochemistry of fungal development. New York: Academic Press 1974
192. Řeháček, Z., Sajdl, P.: Biotechnol. Lett. *1*, 53 (1979)

Induction of Xenobiotic Monooxygenases

Robert V. Smith and Patrick J. Davis
Drug Dynamics Institute, College of Pharmacy
University of Texas at Austin
Austin, TX 78712, USA

The P-450 electron transfer system is a relative young detection of the biochemists described first in liver cells. Only recently has decisive progress been made in microbial cells and in connection with research on utilization of alkane as the sole source of carbon for microbes. Engineers, by studies on the effects of mixing and microdroplet formation, made the initial contributions to our understanding of microbial growth on alkanes. But the understanding of the whole mechanism of uptake was far from being complete. Fundamental studies of alkane uptake and oxidation by yeasts and bacteria soon followed. In Smith and Davis's article one may see that most likely the proper understanding of this extramitochondrial electron transfer chain will help us in future to improve industrial alkane processes, which are so important in some countries. We thus thought that a "purely biological" paper on this important topic should appear in this volume of Advances in Biochemical Engineering. In one of the next volumes two "applied" articles on alkane biology will appear. Hopefully the bridge to the application will then be complete.

The Editor

The characteristics and induction P-450-type monooxygenases are reviewed. A discussion of the location, composition, functional characteristics, and substrate specificities of mammalian liver cytochrome P-450 monooxygenases is provided. This is followed by a description of the effects of inducing agents where particular emphasis is placed on biochemical correlates of induction. After a description of the nature of microbial monooxygenases, induction in these organisms is surveyed. Comparisons are developed for enzymes in fungi, yeasts, and bacteria. Mammalian and microbial hydroxylations of the model hydrocarbon, biphenyl, are reviewed. Results of biphenyl-hydroxylase experiments in mammalian systems are discussed and a rationale described for parallel experiments in microorganisms.

1 Introduction

Eukaryotic organisms are characterized by the presence of cytochrome P-450 monooxygenases which catalyze the oxidation of wide array of organic substrates. *Monooxygenases activate molecular oxygen causing one atom to be incorporated into a substrate and the second reduced and converted into water.* Heme-based monooxygenases are characterized by a spectral band at 450 nm in the visible spectra of their reduced [Fe(II)] carbon monoxide (CO) complexes. These enzymes, hereafter referred to as cytochrome P-450 monooxygenases or just P-450 monooxygenases are different from diooxygenases most commonly found in bacteria, which catalyze concerted incorporation of molecular oxygen into substrates.

Organisms can increase monooxygenase activity to more effectively metabolize certain foreign substances. This response can be qualitative (change substrate selectivity) or quantitative (enhanced capacity to utilize a given substrate or substrates) and will hereafter be referred to as induction.

Induction in mammals was first described in the 1950's by workers at the University of Wisconsin[1, 2] who observed increased metabolism (N-demethylation and azo-reduction) of 3′-methyl-4-dimethylaminoazobenzene in rats following administration of 3-methylcholanthrene (3-MC).

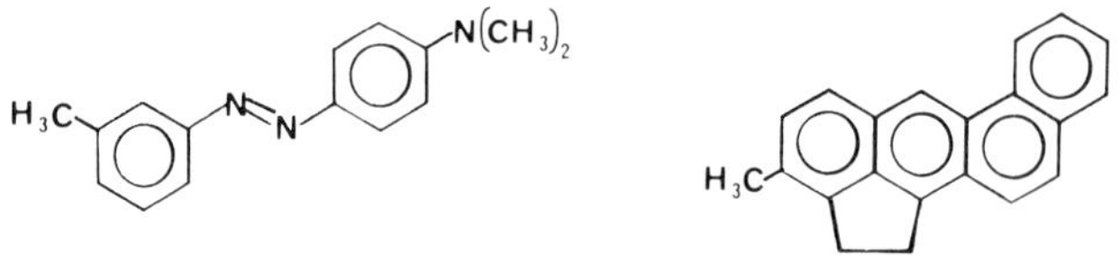

H_3C — $N(CH_3)_2$ — H_3C

3′-Methyl-4-dimethylaminoazobenzene 3-Methylcholanthrene

Since the early fifties, a great many studies have appeared describing the induction of monooxygenases in a variety of eukaryotic organisms.

Induction processes can be of importance in clinical medicine[3, 4].

We have been interested in the comparative metabolism of xenobiotics by mammalian and microbial systems for the past several years. In 1974, Smith and Rosazza described this work for the first time and coined the term "microbial models of mammalian metabolism"[5]. The microbial models approach involves *the use of microorganisms to produce difficult-to-synthesize metabolites* that are formed in only minute quantities by mammalian systems[6–8].

We have become interested in the comparative aspects of induction of P-450-type monooxygenases in mammals and microorganisms. Besides the underlying comparative biochemical importance of this work, we feel that microbial induction studies may lead to methods for altering the course of transformations and improving yields thereby enhancing the versatility of the "microbial models" approach.

2 Characteristics of Mammalian Cytochrome P-450 Monooxygenases

Over the past few years, scientists have debated whether mammalian P-450 monooxygenases are best represented as a single enzyme or a series of enzymes. Investigators wished to reconcile the broad substrate specificity of these enzymes while recognizing that a membrane-bound (stem) enzyme could be fragmented into apparent apoenzymes during solubilization steps. However, considerable evidence now exists supporting the theory that a series of P-450 monooxygenases occur in mammals[9–11]. This evidence as well as discoveries which have further illuminated our understanding of the location, function, and role of mammalian cytochrome P-450 monooxygenases are described below.

2.1 Anatomical Location of P-450 Monooxygenases

As indicated in Table 1, the cytoplasmic network of tubules and vesicles known as the endoplasmic reticulum is the location of most mammalian cytochrome P-450 monooxygenases. These membrane-bound enzymes are primarily associated with the smooth endoplasmic reticulum (SER) which is devoid of ribosomes; the rough endoplasmic reticulum (RER), which is associated with ribosomes, is less important in terms of P-450 monooxygenase content. It is well known that P-450 monooxygenases occur in greatest concentrations in hepatocytes[8, 15]. Thus, in the intact animal, the liver is generally the prime site of xenobiotic transformations. Besides hepatic tissue, P-450 monooxygenases are found in the endoplasmic reticulum of cells in the gastrointestinal tract, lung, and kidney of mammals. Substrate specificity (for steroids) is considerably narrower for kidney P-450 monooxygenase than the other's noted[11]. This is also true for the P-450 monooxygenase derived from adrenal cortex mitochondria[11, 12].

Table 1. Location and relative substrate specificities of cytochrome P-450 monooxygenases

Tissue (cells)	Sub-cellular organelle	Relative substrate specificity	Ref.
Liver (hepatocytes)	SER [a]	Broad [b]	6–8, 10
	Mitochondria	Narrow	13
	Nuclear envelope [c]	–	14
Gastrointestinal tract	SER	Broad	16, 17
Kidney	SER	Narrow	11
	Mitochondria	Narrow	11
Lung	SER	Broad	11
Adrenal cortex	SER	Narrow	18
	Mitochondria	Narrow	11, 12

[a] Smooth endoplasmic reticulum
[b] See Table 3
[c] Not fully defined

The presence of monooxygenases in mitochondria was, for a long time, thought to be an exceptional property of adrenal cortex cells. However, cytochrome P-450 steroid hydroxylating enzymes have recently been derived from rat liver[13)] and chick kidney mitochondria[11)]. A more novel and perhaps more important finding is that cytochrome P-450 monooxygenases are located in the nuclear envelope of rat hepatocytes[14)]. This discovery may have profound toxicological significance particularly in light of the apparent induction of benzo[a]pyrene hydroxylase activity in this enzyme system by 3-MC (see Sect. 3).

2.2 The Nature of Microsomes

When physically disrupted by a process such as homogenization, the SER of cells is converted into microsomes. These proteinaceous-lipoidal spheres can be isolated by a series of steps with ultimate recovery following centrifugation at approximately 100,000 x *g*[19)]. It should be emphasized that microsomes are artifacts of physical manipulation of cells rather than actual anatomical structures. There is occasionally confusion on this point in the literature.

Microsomes are generally thought to be homogeneous, however, subfractionation on sucrose gradients has recently been accomplished indicating that forms of different density are generated upon homogenization of mice liver cells[20)]. Furthermore, density distribution patterns are altered following induction with phenobarbital (Pb)[20)]. While the noted study[20)] suggests different forms of cytochrome P-450 monooxygenases, more definitive evidence for distant enzymes has been gathered by solubilization techniques (i.e., treatment with detergents) and resolution of components *via* column chromatographic and gel electrophoretic methods[11)].

2.3 Composition of Microsomes

Lu and Coon[21] were the first to report the successful solubilization and resolution of hepatic microsomes into three fractions which when recombined, catalyzed the oxidation of fatty acids[21], steroids, drugs, and other xenobiotics[22–24]. In the last few years, the essential components have been identified as cytochrome P-450 monooxygenase(s), NADPH-cytochrome P-450 reductase, and phospholipid[25–30]. Furthermore, characterizing information on these constituents have been obtained as indicated in Table 2. Most noteworthy is the resolution of P-450 monooxygenases from various mammals into as many as ten separate enzymes through sodium dodecylsulfate (SDS) gel electrophoresis[9–11, 31–33].

Table 2. Composition of mammalian hepatic microsomes

Component	Composition	Prosthetic group
Cytochrome P-450 monooxygenase	Series of 4–10 proteins[a], molecular weight range ~ 45,000–65,000	Iron-protoporphyrin IX
NADPH-cytochrome P-450 reductase[b]	Possible hexamer with individual protein components each having a molecular weight of ~ 79,000	FAD and FMN (1:1)
Phospholipid	Phosphotidylcholine and possible other lipids[c]	–

[a] Determined by SDS gel electrophoresis
[b] Probably the same as, or very similar to NADPH-cytochrome c reductase[34]
[c] Can be replaced by dilaurylglyceryl-3-phosphorylcholine or other detergents such as Triton X-100[35]

It is widely believed[9–11] that the different P-450 monooxygenases account for the varying substrate specificities observed with these enzymes. Thus, the different forms of P-450 monooxygenases cannot be regarded as isoenzymes[9] and differences in substrate specificities in various mammals are possibly attributable to differing patterns of these enzymes[9]. Coon et al.[10] leave open the possibility that some substrate specificity may be imparted by multiple forms of NADPH-cytochrome P-450 reductase. However, this question will require further study along with the overall relevance of the above noted findings to man.

Up to this point, we have somewhat loosely referred to liver monooxygenases by the notation, P-450 which we remind the reader, refers to the visible absorption maximum (so-called Soret band) of the reduced CO-complex. However, following the discovery of multiple forms of the cytochrome P-450's and in depth investigation of at least two of these enzymes in pure form, it became apparent that slightly different spectral properties existed for each enzyme.

As will become apparent in later sections, phenobarbital administration to mammals causes a proliferation of cytochrome P-450 monooxygenases possessing a Soret band (for the composite mixture of CO-complexes) at 450 nm. In contrast, certain aryl hydrocarbons (and related substances) such as 3-MC and β-napthoflavone induce the formation of microsomal cytochromes with a Soret band centered at 448 nm. In fact, pure cytochrome P-450 and P-448 have been isolated from rabbits following induction by Pb and β-napthoflavone, respectively. Unfortunately, there are a number of cytochrome P-450 monooxygenases for which no inducing agent has yet been found and this has hampered their isolation and characterization (see Sect. 3). Thus, there is potential for confusion by merely referring to cytochrome P-450 monooxygenases simply by the position of their respective Soret bands.

β-Napthoflavone (5,6-benzoflavone)

As a means of classifying microsomal cytochrome P-450 monooxygenases, Coon et al.[10, 31)] have recommended nomenclature based on mobilities of individual enzymes on SDS-gel electrophoresis. Figure 1 provides a schematic representation of P-450 monooxygenases obtained from rabbit liver microsomes and electrophoretically resolved. At least eight distinct bands are observed which vary in concentration according to the type of inducer used (see additional discussion in Sect. 3). The cytochrome enzymes are numbered 1 through 8 *via* the nomenclature $P\text{-}450_{LM_1}$ to $P\text{-}450_{LM_7}$ (LM = liver microsome) which is based on mobility (the higher the mobility, the larger the numerical notation) and molecular weight (the higher the numerical notation, the greater the molecular

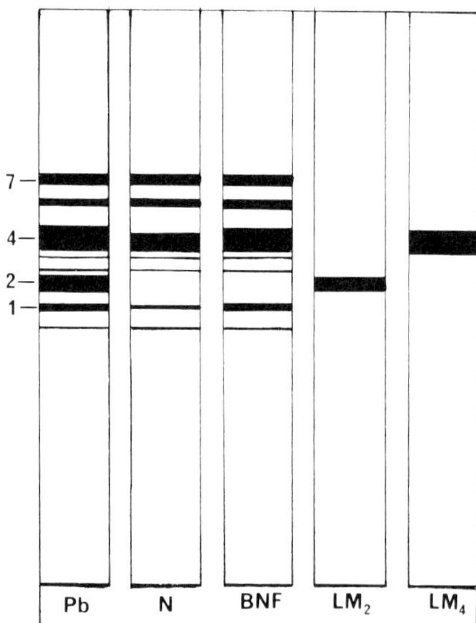

Fig. 1. Diagrammatic representation of rabbit liver microsomal (LM) cytochrome P-450 monooxygenases separated by SDS-polyacrylamide gel electrophoresis; Pb, phenobarbital-induced; N, normal; BNF, β-napthoflavone-induced; LM_2, purified $P\text{-}450_{LM_2}$; LM_4, purified $P\text{-}450_{LM_4}$. (see Ref.[10)])

weight). This represents a convenient index for categorizing cytochrome P-450 monooxygenases and it is our hope that it will be universally adopted; not only for comparisons of monooxygenases from mammals but also for comparative studies with microorganisms (see Sect. 4). Bill and Hodgsen[32)] have recently critically evaluated the experimental routines for the SDS-gel electrophoretic separation of P-450 monooxygenases and their report is recommended to workers who may be unfamiliar with this methodology. The recent isolation and purification of cytochrome P-450 enzymes from rat liver by lauric acid affinity- and gel-chromatography is also noteworthy[32a)].

2.4 Functional Characteristics of P-450 Monooxygenases

Cytochrome P-450 monooxygenases bind substrate and molecular oxygen and interact with reductases in a two-step sequence[33)] to activate the oxygen. Ultimately, one oxygen atom is incorporated into the substrate and the second is reduced to water (Fig. 2). Formation of a P-450-xenobiotic-O_2 ternary complex helps to explain the generally suggested substrate specificity imparted by the cytochrome monooxygenases. The inter-

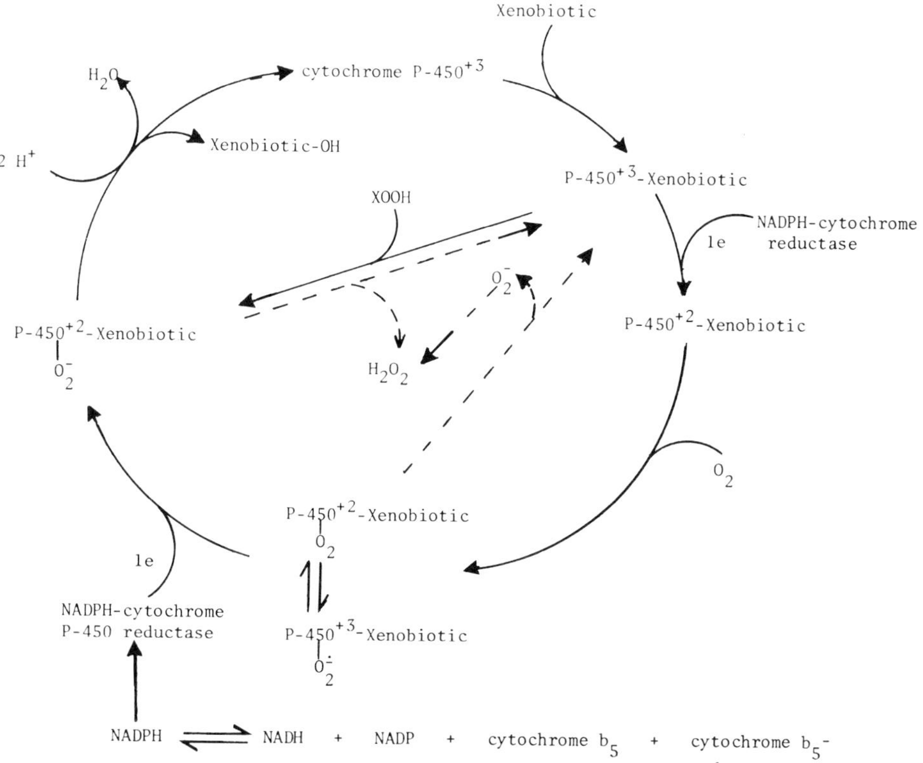

Fig. 2. Oxidation of xenobiotics *via* cytochrome P-450-linked monooxygenases

action of NADPH-cytochrome P-450 reductase in a two-step (2-electron) sequence is also indicated in Fig. 2. Furthermore, the possible coupling of this reductive process with an NADH-mediated electron transfer as shown, explains the often suggested synergistic role for the latter pyridine-nucleotide as well as cytochrome b_5 and cytochrome b_5-reductase[34, 36–38]. A coupling/decoupling role for hydrogen peroxide has recently been proposed[31, 39] as represented in Fig. 2. Thus, NADPH (and NADPH-generating systems) can be replaced by hydrogen peroxide and a variety of organic peroxides in the oxidation of certain xenobiotics by microsomal mixtures[40–45] and cytochrome P-450_{LM_2}[46] in particular. This is an intriguing finding and may support the postulated intermediacy of the P-450^{+2}-O_2^- complex depicted in Fig. 2. However, the role of hydrogen peroxide in intact hepatocytes may be of lesser importance[46a].

Cytochrome P-450 monooxygenases are differentially inhibited by a variety of substances including CO, SKF-525A, metyrapone, ethylisocyanate, and the Lilly compound, 2,4-dichloro-6-phenylphenoxyethylamine (DPEA)[47–51c]. Indeed, varying inhibitory effects on different microsomal preparations are suggested as underlying support for multiple forms of P-450 monooxygenases[11]. However, inhibitor studies must be interpreted with caution, especially with heterogenous preparations. For example, microsomal hydroxylase activity is widely thought to be insensitive to cyanide[52–54]. However, cytochrome P-450 systems have recently been found to be inhibited by cyanide in a concentration-dependent manner[55, 56].

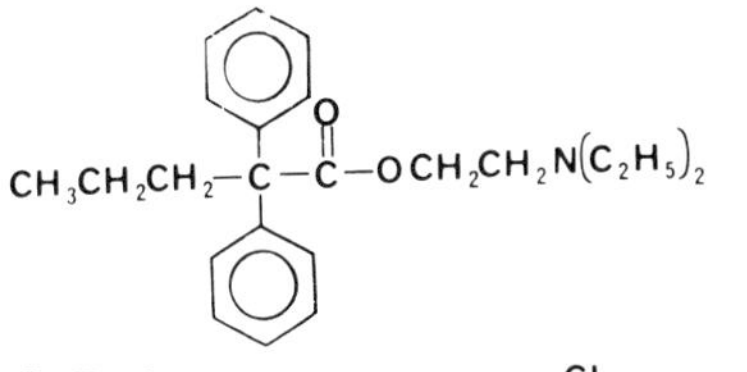

SKF-525 A

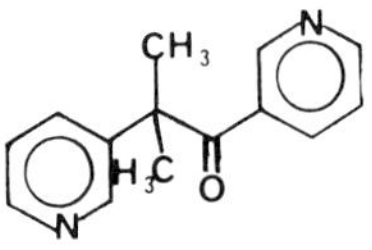

Metyrapone

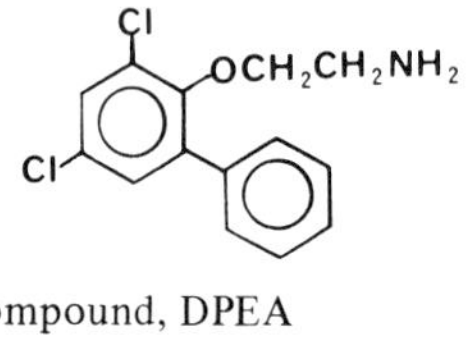

Lilly Compound, DPEA

2.5 Substrate Specificity

Cytochrome P-450 monooxygenases catalyze a wide variety of oxidations with a vast number of substrate-types as indicated in Table 3. Indeed, many authorities suggest[9–11] that this fact supports the purported multiplicity of P-450 monooxygenases. It is thought that few (and perhaps no liver microsomal) P-450 monooxygenases (e.g., P-450_{LM_2}) catalyzes a single reaction; but certain cytochromes will have greater activity of a certain type than others. For example, cytochrome P-448 (LM_4 from rabbits) possesses much greater aryl hydroxylase activity than cytochrome P-450 (LM_2 from rabbits).

Table 3. Examples of oxidative transformations catalyzed by mammalian cytochrome P-450 monooxygenases

Reaction type	Representative substrate(s)	Products	Ref.
Aliphatic hydroxylation	Octane, dodecane, cyclohexane	1° and/or 2° alcohols	23
Aromatic hydroxylation	Acetanilide, benzo[a]pyrene, chlorobenzene	Arene oxides, phenols	57–59
N-Dealkylation	Benzphetamine, ethylmorphine	2° Amine + formaldehyde	23, 60
O-Dealkylation	Ethoxyresorufin, ethylmorphine	Phenol + acetaldehyde	60–62
Desulfuration	Parathion	Oxoester	27, 63
N-Hydroxylation	2-Acetaminofluorene	N-Hydroxy-acetamino-fluorene	64

3 Induction of Mammalian Cytochrome P-450 Monooxygenases

The stimulation of liver growth in mammals by exposure to foreign organic compounds, has been known for more than three decades[65]. In the last ten years a number of insights have been gained on induction and its molecular correlates. Indeed, the use of inducing agents to produce microsomal fractions enriched in certain P-450 monooxygenases has been of immeasurable benefit in the isolation and purification of the few forms of P-450 (e.g., LM_2 and LM_4 from rabbits) that have been subject to characterization. There is reason to believe that a more thorough understanding of induction processes and the P-450 monooxygenases *per se* will develop as studies continue with purified preparations.

3.1 Gross Effects

The stimulation of tissue growth by inducing agents can be described as individual cellular (e.g., hepatocyte) growth or cell proliferation. These events are described by the terms hypertrophy and hyperplasia, respectively. However, it is noted repeatedly in the literature that hyperplasia should also include proliferation of genetic material[65]. The hypertrophy observed with inducing agents is related to increased *de novo* synthesis of proteins, which can serve as precursors to cytochrome P-450 monooxygenases[65a]. Thus, animals appropriately treated with inducing agents such as Pb or 3-MC will show a proliferation of SER. The resulting increases in microsomal yields and their enzymatic activities are discussed below.

3.2 Inducing Agents

Inducers of hepatic P-450 monooxygenases consist of compounds possessing a broad spectrum of structural types, uses, and pharmacological activities. A representative listing of mammalian inducing agents is given in Table 4. The only features that these compounds appear to share are:

1. high lipid-solubility permitting ready localization in the endoplasmic reticulum, and

2. ability to serve as substrates for, or the capacity to bind to P-450 monooxygenases[66)].

Following administration of an inducing agent, different effects are observed depending on the particular nature of the agent. Gillette[67, 68)] classified materials into three groups, depending on their action as inducers. First, phenobarbital and many other drugs increase levels of cytochrome P-450 and NADPH-cytochrome c reductase. Secondly, 3-methylcholanthrene and other polycyclic hydrocarbons increase cytochrome P-448 but not levels of P-450, NADPH-cytochrome c reductase, or the rate of P-450 reduction. Thirdly, spironolactone and other steroids increase NADPH-cytochrome c reductase and the rate of P-450 reduction and may[66)] effect levels of cytochrome P-450. These steroidal inducers are much less commonly used experimentally but may be of particular importance since some are natural substrates for monooxygenases. The Gillette classification system contains some useful generalities; however, it does point out a problem of nomenclature. Much of the early literature and even some recent literature indiscriminately employs the term "cytochrome P-450" to mean all forms of the cytochrome. In many of the more recent publications the terms cytochrome P-448 and cytochrome P-450 represent specific forms having absorbance maxima for their reduced CO-complexes at 448 nm, respectively. These papers contain an inherent ambiguity since P-448 (also called P_1-450 and P-450$_{LM_4}$) and P-450 referred to are probably mixtures of several cytochrome enzymes. As we pointed out earlier, SDS gel electrophoresis of solubilized microsomes from uninduced animals reveals several cytochrome P-450 apoproteins[10, 31, 69–72)]. Inducers cause increases in the levels of one or more of these apoprotein bands[73)]. Tetrachlorodibenzodioxin (TCDD) and 3-MC increase a single so-called "band 4" of mouse liver microsomes (probably P-448 and perhaps analogous to LM_4 described in rabbits[10)]) while Pb affects another[74)]. However, there is some "cross-induction"; for example, Pb in mice also causes slight increases in "band 4". Additionally, recent evidence suggests that "band 4" may not be homogeneous[73)]. Some workers report that cytochrome P-450 from uninduced rats is different than the P-450 produced in Pb-treated animals[75)]. This is probably due to the induction of a form of P-450 that is present in low concentration in uninduced animals. An analogous situation is observed in rabbits where LM_2 markedly increases in rabbit liver microsomal preparations following Pb-induction[10)]. In future years, it is our hope that a standardized SDS gel electrophoresis procedure will be adopted so that cytochrome P-450 monooxygenases from various species and resulting after a number of induction treatments, can be compared. Such a procedure would also help in comparative induction studies with mammalian and microbial microsomal preparations.

Table 4. Examples of agents that affect induction of hepatic cytochrome P-450 monooxygenases

Inducing agent	Description (use and/or pharmacological effect)	Structure
Benzo[a]pyrene	Carcinogen	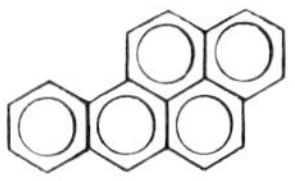
Butylated hydroxytoluene	Antioxidant	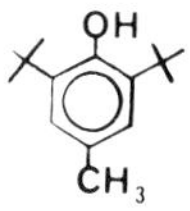
3-Methylcholanthrene	Carcinogen	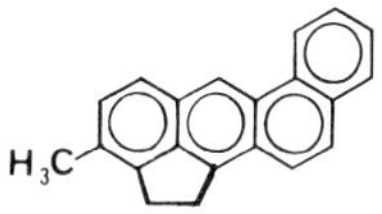
β-Napthoflavone	Hydrocarbon analog	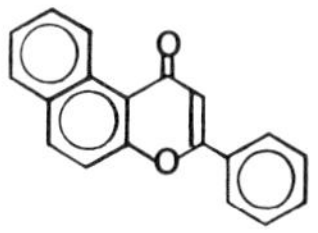
Phenobarbital	Sedative, hypnotic	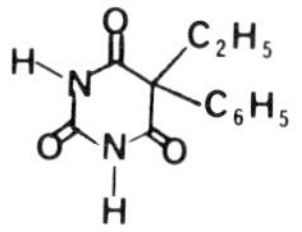
Polychlorinated biphenyls (arochlors)	Heat-exchange fluids, insulators, lubricants, and plasticizers	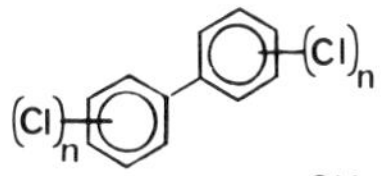
Pregnenolone 16 α-carbonitrile	Steroidal derivative	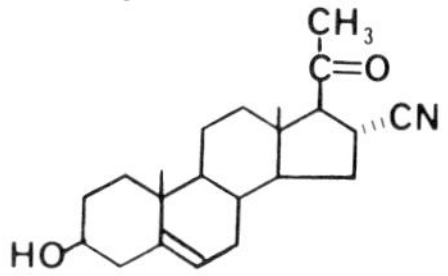
Safrole	Carcinogen, formerly a flavoring agent	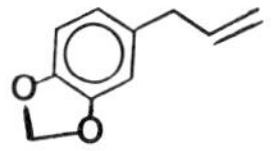
2,3,7,8-Tetrachlorodibenzo-*p*-dioxin (TCDD)	Herbicide impurity; teratogen	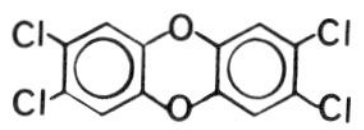
1,1,1-Trichloro-2,2-(*p*-chlorophenyl) ethene (*p,p'*-DDT) and other isomers	Insecticide	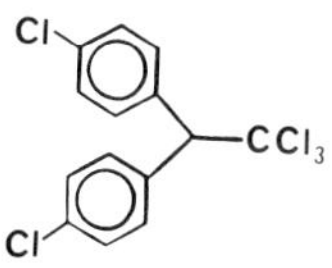

Several workers have recently employed EPR to study induction processes and resulting effects on the spin state(s) of the iron atom of P-450 monooxygenases[31, 76]. Low- and high-spin form of P-450 have been detected; the proportion of each changes with different inducing agents. Interestingly, distinctly different EPR spectra have been for rabbit liver P-450$_{LM_2}$ (Pb-induction) and P-450$_{LM_4}$ (3-MC induction). This additional tool for distinguishing P-450 monooxygenases is likely to become more useful especially in possibly delineating binding mechanisms associated with P-450-substrate interactions.

Differences in substrate specificity have been utilized by a number of investigators to categorize inductions of cytochrome P-450 monooxygenases. Correlations are partially successful as can be seen from Table 5. Pb-induction in rats has a pronounced effect on pentobarbital hydroxylation[75], benzphetamine N-demethylation[75], and aminopyrine N-demethylation[80] that is not observed to any significant extent for microsomal preparations from rats induced with 3-MC.

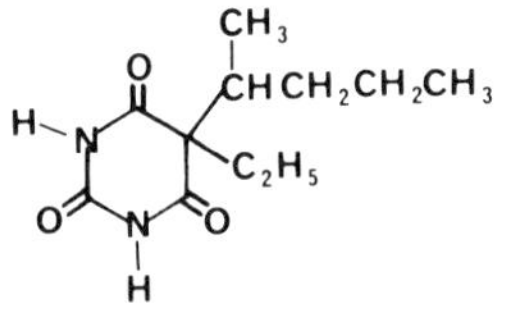

Pentobarbital

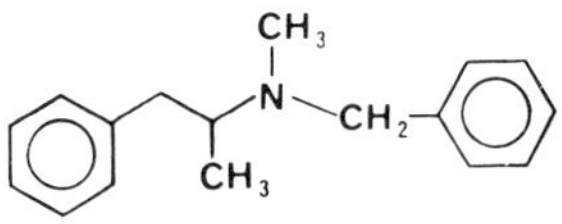

Benzyphetamine

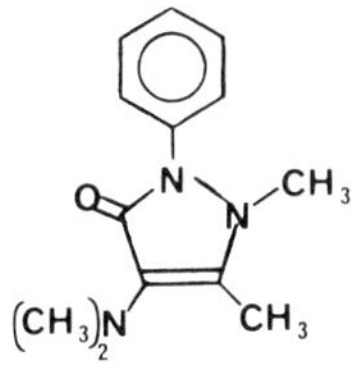

Aminopyrine

Table 5. Substrate specificities of liver microsomal preparations obtained from rats pretreated with phenobarbital and 3-methylcholanthrene

Substrate	Transformation	Relative increase in specific activities[a]	
		Pb-treated	3-MC-treated
Pentobarbital	Aliphatic hydroxylation	1,000	60
Benzphetamine	N-Dealkylation	250	50
Ethylmorphine	N-Dealkylation[b, c]	100 (?)	100
	O-Dealkylation[a, d]	100	100
Aminopyrine	N-Demethylation[e]	480	100
Aniline	Aromatic hydroxylation	120	170
Benzo[a]pyrene	Aromatic hydroxylation	70	390
Testosterone	6β-Hydroxylation	100	100
	7α-Hydroxylation	100	200
	16α-Hydroxylation	550	50

In contrast, 3-MC-induction has a much greater influence on benzo[a]pyrene hydroxylation[75]. For this reason, the cytochrome principally induced by 3-MC is often referred to as aryl hydrocarbon hydroxylase. The reader should also recall that this monooxygenase is synonomous with P-448 or P_1-450. Some anomalies exist in categorizing effects of Pb and 3-MC according to substrate specificities. For example, O-dealkylase activity towards 7-ethoxycoumarin[81], *p*-nitroanisole[81], and ethoxyresorufin[62] is induced by 3-MC but not by Pb. However, as indicated in Table 5, O-dealkylation of ethylmorphine (a substrate widely used as a model in mammalian transformation experiments) is neither effected by Pb or 3-MC.

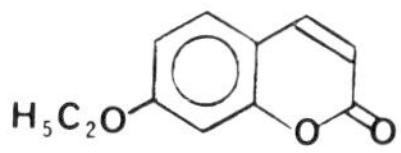

H_5C_2O

7-Ethoxycoumarin

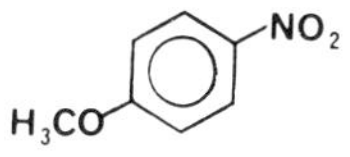

NO_2 H_3CO

p-Nitroanisole

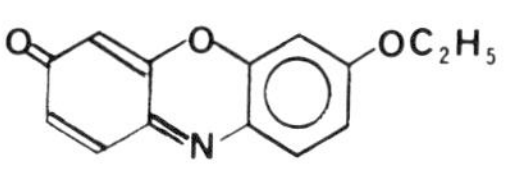

OC_2H_5

Ethoxyresorufin

$N-CH_3$ H_5C_2O OH

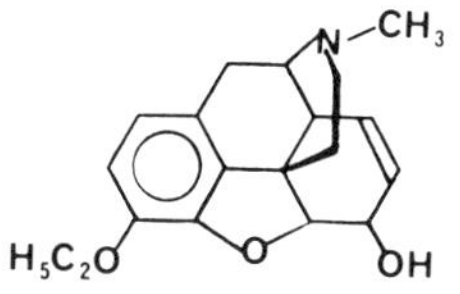

Ethylmorphine

Overall, the data in Table 5 emphasizes the problem of developing patterns of substrate specificity based on results from relatively "crude" microsomal preparations. More work is required on transformations by P-450 monooxygenases. Also, studies should be initiated with closely related series of compounds to detect more subtle differences that may exist in the substrate specificity of different cytochrome monooxygenases.

Recent studies of the induction of hepatic monooxygenases by the polychlorinated biphenyls (Table 4) have led to some interesting structure-activity relationships. Aroclor 1254 is a mixture of principally polychlorinated biphenyls containing 54% chlorine by weight. This relatively highly chlorinated mixture of compounds is reported to be an effective inducer of cytochrome P-450 monooxygenases. In rats, the pattern of induction mimics that observed by combined administration of Pb and 3-MC[82, 83]. Indeed, Ryan et al.[84] isolated two P-450 monooxygenases from liver microsomes of rats treated with the Aroclor 1254 mixture. One of the cytochromes was electrophoretically and antigenetically identical with the principal cytochrome monooxygenases (P-450)

---

a Activities of 100 assigned to untreated controls; data from Ref.[75] unless otherwise indicated

b Ref.[77, 88]

c Nerland and Mannering[79] note that the level of N-dealkylation is increased by Pb-induction but site unpublished results

d Ref.[79]

e Ref.[80]

obtained from the livers of Pb-induced rats. The second appeared to be essentially identical with that obtained from 3-MC-induced rats (P-448).

Poland and Glover[85] systematically studied the induction properties of a number of components of Aroclor 1254 under the assumption that certain constituents were responsible for the two types of induction (P-448). In this work, the authors took advantage of two substrate specificities mentioned earlier. That is, the marked increase in aminopyrine N-demethylase after Pb-induction and benzo[a]pyrene hydroxylase following 3-MC-induction. Thus, they assumed that compounds disproportionately increasing aminopyrine N-demethylase were more like Pb in their effect while congeners (or isomers) showing greater aryl hydroxylase activity behaved more like 3-MC. Sixteen variously halogenated biphenyls (chlorinated and brominated) were evaluated. Compounds increasing benzo[a]pyrene hydroxylase were all highly substituted, lipophilic, and relatively planar. One such compound is 3,4,5,3′,4′,5′-hexachlorobiphenyl.

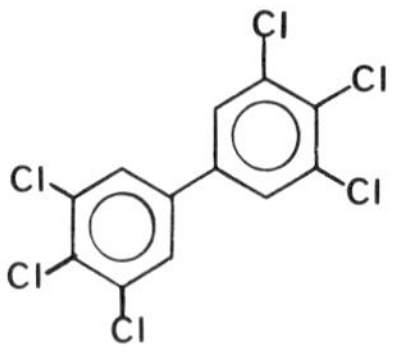

3,4,5,3′,4′,5′-Hexachlorobiphenyl

In contrast, distinctly non-planar biphenyls such as those with chlorosubstituents in the 2- and 2′-positions behaved more like Pb in mice and rats[85]. The planar biphenyls induced a cytochrome monooxygenase possessing a Soret band at 448 nm (reduced CO-complex) while the corresponding cytochrome-CO complex obtained from microsomes of animals induced with non-planar polychlorinated biphenyls showed an absorption maximum at 450 nm. These results support the work of Ryan et al.[84]. Furthermore, Poland and his co-workers provided additional insight into the induction behavior of several other chlorinated-planar substances such as TCDD (Table 4)[85], 2,3,6,7-tetrachlorodibenzofuran[85], and 2,3,7,8-tetrachlorodibenzofuran[86,87].

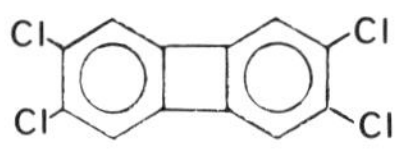

2,3,6,7-Tetrachlorobiphenylene

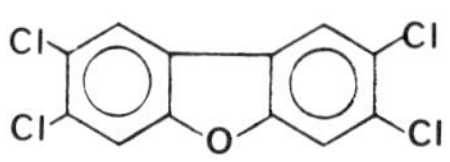

2,3,7,8-Tetrachlorodibenzofuran

These compounds are powerful inducers of cytochrome P-448. Interestingly, it has been observed that the relative activities of the P-448 inducers indicated above can be correlated to their relative affinities for a cytosol protein present in the liver of rats and mice[85,86]. This finding may represent an important first step in a number of induction processes and deserves further study.

Besides the direct indication of multiple P-450 enzymes afforded by the actual isolation of the various enzymes themselves, other indirect evidence may be gained by use of suitable probes of enzymatic catalysis. Where a substrate yields a variety of products, changes in product profiles with varying conditions (inducers, inhibitors) may indicate that these products are not produced by the same enzyme(s). Trager[87a] has described the ingenious use of warfarin as such a probe, based on both the stereospecificity of metabolism (R-*vs.* S-Warfarin) and the regiospecificity of hydroxylation.

The latter potential arises because hydroxylation occurs at the 6-, 7-, 8-, 4′- and benzylic positions

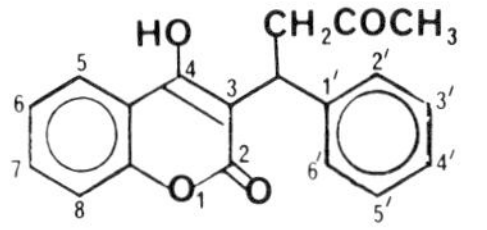

Warfarin

By careful examination of the Km values of the various hydroxylations, and the product ratios of induced *vs.* uninduced animals, conclusions were drawn as to the multiplicity of enzymes involved. For example, the 6-hydroxylation of R-Warfarin showed a product ratio of phenobarbital induced to uninduced (central) of 1.96 ± 0.15, a value comparable to that observed for 7-, 8-, and 4′-hydroxylation, indicating a common hydroxylase. However, differential induction was seen for benzylic hydroxylation which showed a product ratio of 1.5 ± 0.18 thereby implying the involvement of a separate hydroxylase. Similar comparisons of statistically differential inductions of product formation in R- and S-warfarin indicated that at least five distinct enzymes were responsible for formation of the ten hydroxylated products; one for R-benzylic and one for S-benzylic hydroxylation; one for S-7- and one for S-8-hydroxylation; and one for the production of the other six hydroxylated products. This type of study appears to be a sensitive probe of hydroxylase activity and will probably expand as a tool for examining multiple P-450 enzymes, although it is obviously not limited to these.

Diet can influence the course of induction of hepatic monooxygenases[88]. A number of natural foods contain inducing agents. These include members of the *Brassicacae* family (e.g., radishes) as well as spinach, turnips, celery, and dill. The naturally occurring inducers in these foodstuffs are probably terpenes. Furthermore, rats are known to be induced by cedarwood bedding[89], aerosols of eucalyptol[90], and a variety of other terpenes[91, 92].

3.3 Mechanisms of Induction

Up to this point, induction has been described in terms of increased production of cytochrome P-450 monooxygenases. It is generally thought that this proliferation of cytochromes occurs *via* stimulation of protein synthesis. There is much evidence to support this hypothesis[93] and recent investigations have uncovered a number of loci that may be important in the mechanism(s) of induction.

Various interactions that could ultimately lead to increased protein synthesis are noted in Fig. 3. Most important of these are the binding of inducers such as 3-MC to nuclear DNA. This stimulates expression of messenger RNA's which play an integral role in ribosomal activity[93]. The inducer-DNA interaction may be potentiated by covalent binding and important active metabolites could be formed through recently discovered nuclear cytochrome P-450 monooxygenases[94–98]. Other effects such as the direct influences of inducers on ribosomal synthesis and stabilization (*via* ribonuclease inhibition), and protein initiation, elongation, and termination factors have been excellently reviewed by Bresnick[93] and do not require reiteration. In summary, a number of known inducers appear to work at several levels of RNA and protein synthesis to increase levels of P-450 monooxygenases.

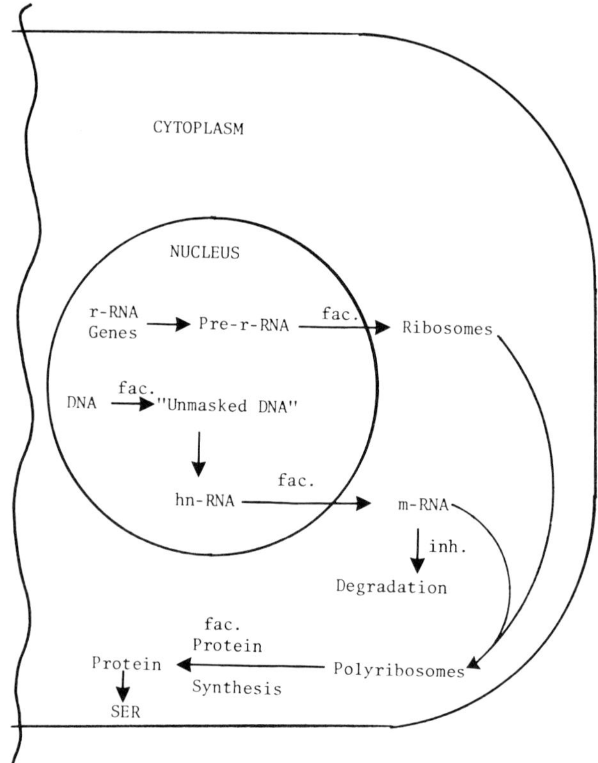

Fig. 3. Sites within mammalian cells that are reportedly important in the induction of cytochrome P-450 monooxygenases. Inducers may facilitate (fac.) or inhibit (inh.) processes as indicated (see Ref.[93]). Abbreviations: hn-RNA (heterogeneous nuclear RNA), m-RNA (messenger RNA), r-RNA (ribosomal RNA), SER (smooth endoplasmic reticulum)

4 Characteristics of Microbial Monooxygenases

4.1 Fungi

Ambike et al.[99] demonstrated the presence of a cytochrome P-450 monooxygenase in the microsomal fraction of a *Claviceps purpurea* strain that produces clavine-type alkaloids. Time course studies showed a *direct correlation between alkaloid production and P-450 monooxygenase levels,* thereby indicating an *in vivo* function of the enzyme. Additionally, cytochrome b_5 was detected in the *C. purpurea* studied. It is interesting to note that cyanide inhibited the production of cytochrome P-450 in this organism while the total alkaloidal yield decreased proportionately to the levels of the enzyme; hence, the P-450 monooxygenase was not inhibited *per se* by cyanide. The interconversion of cytochrome P-450 and cytochrome P-420 was examined in this system[100]. Following removal of cytochrome b_5 by steapsin (a lipase), the conversion of cytochrome P-450 to cytochrome P-420 was effected by snake venum protease, sodium cholate or oxygen. The reconversion of P-420 could be accomplished in quantitative yield by treatment with 10% glycerol, which also prevents the P-450 to P-420 transformation by steapsin. A similar transformation (P-420 to P-450) has been reported with mammalian monooxygenases using polyols or reduced glutathione[101] which presumably act by uncoupling detergents and/or sulfhydryl reagents from cytochrome P-420.

The exact function of the P-450 monooxygenase described by Ambike et al.[99, 100] is unknown. However, it would be interesting if it behaved as a hydroxylase in the conversion of tryptophan to chanoclavine because of the mechanistic implications of the presumed regiospecific hydroxylation necessary for the biosynthesis of the latter[102].

In a preliminary communication, Auret et al.[103] described the now well-defined NIH-shift during the *aromatic hydroxylation of anisole* by nine fungal species. The NIH-shift, which has been suggested to be *a priori* evidence for arene oxide formation by monooxygenases (in conjunction with other evidence[104, 105]) involves a rearrangement of a substituent, for example at the *para*-position, during the spontaneous rearrangement of the epoxide to the phenol as indicated in Fig. 4. In a study expanding on these results[106], the amount of deuterium retention was shown to be ~ 70%, which is comparable to values obtained with *in vitro* mammalian liver systems during *p*-hydroxylation of anisole. Comparable retentions were also observed using, 3,5-di-^{2}H-anisole indicating that the initial position of the deuterium atom does not influence the amount of retention, consistent with a 3,4-oxide and the NIH-shift (see Fig. 5).

An interesting observation reported by Boyd et al.[106] deserves comment. Young cultures, which presumably consisted of spores and germinating spores, showed a preponderance of *O-demethylation of anisole to phenol,* while aromatic hydroxylation was observed in older cultures. An investigation of this phenomena is certainly warranted both from a fundamental biochemical standpoint (different enzymes or a single, altered enzyme), and because of the current widespread interest in using microbial cultures to carry out specific type-reactions including O- and N-dealkylations and aromatic hydroxylation[107–110].

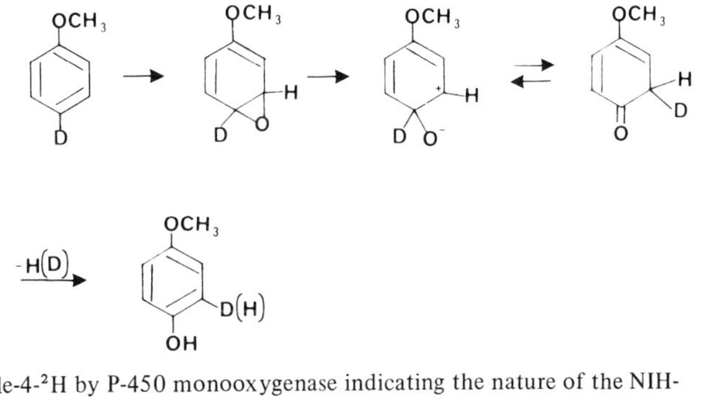

Fig. 4. Hydroxylation of anisole-4-^{2}H by P-450 monooxygenase indicating the nature of the NIH-shift

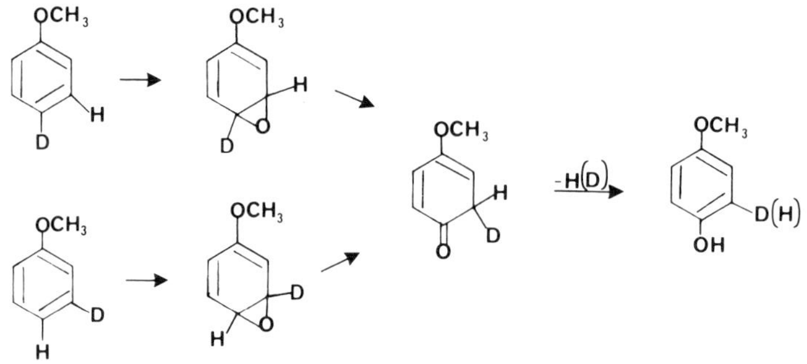

Fig. 5. Mechanism for equivalent retention of deuterium during the aromatic hydroxylation of anisole-3-^{2}H and anisole-4-^{2}H by cytochrome P-450 monooxygenase

Unlike mammalian species, which show a preference for *p*-hydroxylation of anisole, most of the fungal species tested showed a predominance of *o*-hydroxylation. It must be remembered, however, that this is with a single substrate, and a limited number of fungi. Deuterium studies concerning the *o*-hydroxylation of anisole were consistent with a 2,3-epoxide rather than a 1,2-epoxide, although lower deuterium-retention values were obtained than with liver microsomes. The low values of deuterium retention in the *o*-hydroxylation of aryl acids with *Aspergillus niger* prompted the proposal of a different mechanism involving a 1,2-oxide, or an insertion mechanism[103, 106]. The lack of an observed isotope effect would tend to discount, but not completely disprove the latter hypothesis.

Two studies by Ferris et al.[111, 112] focused on a fungal monooxygenase system capable of carrying out *reactions similar to those of mammalian liver systems. Cunninghamella bainieri* (ATCC 9244) was chosen as the model system which was used as a cell suspension supplemented with NADPH. In analogy to mammalian work, this culture was shown to carry out the following metabolic reactions: N-demethylation of aminopyrine; O-demethylation of *p*-nitroanisole and anisole; hydroxylation of anisole, aniline, and naphthalene; and reduction of both 1,2-dimethyl-4-(*p*-carboxyphenylazo)-5-hydroxybenzene and *p*-nitrobenzoic acid. This diversity in the ability to metabolize a number of xenobiotics is of prime interest in considering that microbial systems mimic the types of reactions, and indeed, can be used as models for mammalian metabolism is proposed by Smith and Rosazza[6, 7].

In analogy to previously cited work[103, 106], Ferris et al.[111, 112] reported evidence for the NIH shift in the *p*-hydroxylation of anisole, with the expected degree of deuterium retention. No primary isotope effect was observed as would be expected if an insertion mechanism were operating. Isotopic labelling using $^{18}O_2$ revealed incorporation of a single atom of ^{18}O which is indicative of a typical monooxygenase. Interestingly, *Cunninghamella bainieri* (ATCC 9244) predominantyl caused *p*-hydroxylation, while *C. elegans* (ATCC 9245)[103, 106], principally effects *o*-hydroxylation of anisole. Naphthalene was metabolized to the *trans*-dihydrodiol, indicating the presence of exposide hydratase, and to a 9: 1 mixture of 1- and 2-naphthol in similar fashion to mammals.

Conclusive evidence was presented for a cytochrome P-450 in *C. bainieri*[112]. A typical difference spectrum of the reduced P-450-CO complex indicated a peak at 450 nm, after removal by digitonin solubilization of a strongly negatively absorbing species (445 nm). Hydroxylase activity, as determined by a benzo[a]pyrene assay, was found to be considerably less than that of rat liver, although the assay was not performed at the pH optimum and this was indicated by the authors[112]. The aryl hydrocarbon hydrolase activity was found to be NADPH dependent as with mammalian microsomal P-450, and was also inhibited by SKF-525 A, metyrapone, and carbon monoxide.

Microorganisms have been used for decades in *commercial processes involving the hydroxylation of steroids.* Nearly every position on the carbocyclic skeleton can be hydroxylated by the proper choice of cultures[107, 113]. Recently, Breskvar and Hudnik-Plevnik[114, 115] have implicated a cytochrome P-450 in the 11-α-hydroxylation of progesterone by *Rhizopus nigricans.* The hydroxylation was shown to occur in both cell suspension and cell-free systems. The reaction was inhibited by CO, and light reversed the inhibition. The difference spectrum of the membrane fraction sedimenting at 105,000 x *g* showed a peak at 450 nm, indicative of the reduced P-450-CO complex. This cytochrome was not found in mitochondria, but did sediment at high g-values indicating a membrane bound, rather than soluble cytochrome P-450. The substrate, progesterone, exhibited a typical type II binding with cytochrome P-450, as defined by Shenkman et al.[116].

One of us, in conjunction with J.P. Rosazza, examined a selected group of microorganisms, mostly fungi, to carry out *aryl hydroxylations* similar to those reported for mammalian systems[5]. Many of these cultures were chosen based on their reported ability to carry out this type-reaction, and indeed, some had been shown to contain cytochrome P-450. The metabolism of a number of model aromatic compounds were examined, including acetanilide, aniline, anisole, benzene, benzoic acid, biphenyl, chlorobenzene, coumarin, naphthalene, nitrobenzene, *trans*-stilbene, and toluene. Generally, the results obtained showed good qualitative comparisons to those reported in mammalian systems. A closer parallel was also noted in a number of cases, to *in vitro* rather than *in vivo* mammalian results.

Smith, Rosazza et al.[117] also examined the *aromatic hydroxylation of isomeric xylenes* using the cultures noted above[5]. Generally a close correlation was found between the positions of hydroxylations by microbial and mammalian species. These results, as well as parallels reported by others as already discussed, prompted Smith and Rosazza[6, 7] to suggest that it may be possible to define microbial transformation systems that could mimic many of the biotransformations occurring in mammals. This area of study was dubbed, "microbial models of mammalian metabolism". Ample literature is available to facilitate choosing microorganisms to carry out desired reactions[106–110]. The same advantages of scale-up would apply to the isolation of enzymes for further study and commercial use, such as in cytochrome P-450 production. The future prospects of such production were recently reviewed by Wiseman[118].

A cytochrome P-450 has been implicated by Murphy et al.[119], in the *biosynthesis of patulin* by *Penicillium patulum* (see Fig. 6). The hydroxylation of *m*-cresol to *m*-hydroxybenzyl alcohol and 2,5-dihydroxytoluene was shown to require NADPH and mo-

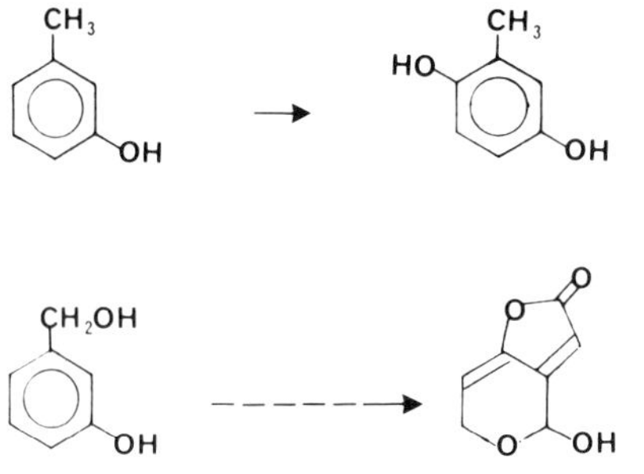

Fig. 6. Steps in the biosynthesis of patulin by *Penicillium patulum* (see Ref.[119])

lecular oxygen. The conversions are possibly conducted by different enzymes in separable membrane fractions. The formation of both products was strongly inhibited by CO, but only marginally by cyanide. The CO inhibition was reversed by light, and evaluation of various wavelengths indicated maximal reversal at 450–455 nm. Only low levels of a cytochrome P-450 were observed in the reduced-CO difference spectrum, and a peak at 420 nm was always present. All attempts at solubilizing the apparently particulate monooxygenase were unsuccessful.

Cerniglia and Gibson[120] examined the ability of a *Cunninghamella elegans* strain, isolated by enrichment culture on crude oil, to *metabolize naphthalene.* These authors found close correlation to the routes of naphthalene metabolism in mammalian systems. Major metabolites included 1-naphthol and 4-hydroxy-1-tetralone. Four other minor metabolites were also isolated and characterized. In a second study, the microsomal preparation from that fungal strain was found to oxidize napthalene in a typical monooxygenase fashion[121]. Both 1- and 2-naphthol, as well as the *trans*-dihydrodial were found. These compounds are indicative of a common 1,2-epoxide intermediate, as well as the presence of epoxide hydratase which has been seen in at least one other fungal species[111, 112]. Metabolism was dependent upon NADPH, and O_2, and was also inhibited by CO but not cyanide. Studies with generally accepted P-450 inhibitors, such as SKF-525A and metyrapone, were also consistent with this view. A typical spectrum of the reduced P-450-CO complex was obtained indicating a cytochrome P-450 with higher levels of cytochrome P-420.

An interesting proposal presented[120] was that *1,4-oxygenated products of napthalene* may result from the formation of a naphthalene-1,4-endoperoxide, presumably by a dioxygenase. This suggestion has been made previously for naphthalene by Schäfer-Ridder et al.[122], and a similar result has actually been demonstrated in the microsomal transannular 1,4-peroxidation of 9,10-dimethyl-1,2-benzanthracene to give the 9,10-epidioxide which is then reduced to the *cis*-diol[123]. Such a metabolic reaction would propably involve a dioxygenase and, hence, $^{18}O_2$ labeling would be definitive in that regard. The reaction may also be thought of, in a chemical sense, as a Diels-Alder addition of molecular oxygen. Further implication of enzymatic involvement in such a process would be the chiral specificity that one might expect with the resultant epidioxide of the benzanthracene, or of a substituted naphthalene (see Fig. 7). It is of interest that similar metabolism studies to those previously cited[120] are being conducted with bac-

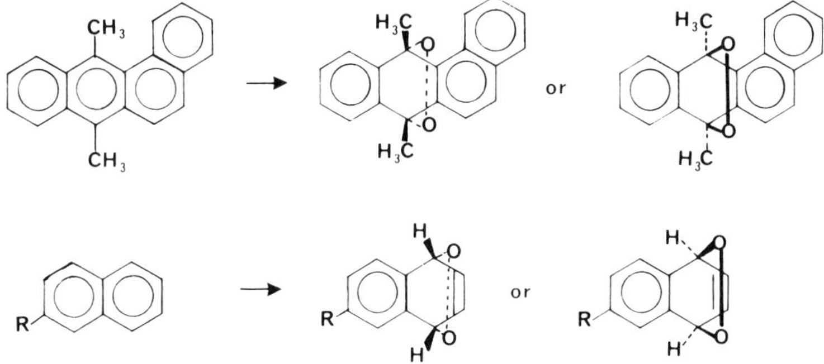

Fig. 7. Possible products resulting from diooxygenase-catalyzed oxidation of 9,10-dimethyl-1,2-benzanthracene and a substituted naphthalene

teria, however, sufficient data is not yet available to implicate P-450 monooxygenase involvement in their metabolism[124].

Although numerous fungal species are able to utilize alkanes for growth, studies have only recently addressed the possible involvement of monooxygenases in the *metabolism of gaseous alkanes.* An *Acremonium* strain (BE/1) was isolated based on its ability to utilize ethane, propane, and butane[125]. These were shown to be metabolized by the alcohol-aldehyde-acid route. While a difference spectrum of P-450 could not be demonstrated in a microsomal preparation, inhibition by CO but not cyanide and dependence on NADPH was observed which are strongly suggestive of a monooxygenase, in analogy to those found in several yeasts (see below).

4.2 Yeasts

A monooxygenase system from *Torulopsis sp.* capable of *hydroxylating the penultimate carbon of fatty acids* has been described by Heinz et al.[126]. Activity was shown both in the intact cell, and in a cell-free system consisting of the 10,000 x *g* supernatant fraction (microsomes). The enzyme(s) was found to be stabilized by glycerol, but not by sulfhydryl compounds (dithiothreitol, mercaptoethanol) or ascorbate. Studies indirectly implicated a P-450 type monooxygenase by the incorporation of one oxygen atom from $^{18}O_2$, and the requirement for NADPH. Additionally, the hydroxylation of fatty acids was inhibited by CO, and not by cyanide.

Lebeault et al.[127] demonstrated the presence of a cytochrome P-450 in *Candida tropicalis* utilizing *tetradecane* as a sole carbon source. This 27,000 x *g* supernatant fraction (containing microsomes) was capable of affecting ω-hydroxylation of fatty acids and alkanes, the most active substrate being laurate. A similar system may be involved in alkane metabolism by *Candida intermedia* (NRRL-Y6328-1)[127a]. This system also showed a dependence on molecular oxygen and NADPH, and was inhibited by CO but not cyanide. A typical difference spectrum indicative of a cytochrome P-450 was observed, but only in yeasts grown on hydrocarbon. In addition, the system was capable of N-dealkylating a number of drug substances including aminopyrine, benzphetamine, and ethylmorphine[127].

In a second paper, these investigators examined the partial purification of the P-450-like system and the role of phospholipids[128]. Interestingly, the cytochrome P-450 prepared by fractionation on DEAE cellulose was not readily auto-oxidized, and could be further stabilized by glycerol or dithiothreitol. Such observations are encouraging for the preparative use of such systems as described by Wiseman[118]. Other fractions included cytochrome reductase and a heat stable lipid, both of which were required with the cytochrome P-450 for activity. In evaluating the effects of phospholipids on the ω-hydroxylation of lauric acid, it was found that yeast lysophosphatidylethanolamine gave highest activity with the reconstituted P-450 system. This system was also active in catalyzing the N-dealkylation of the drug substances noted above, but was not capable of O-demethylating *p*-nitroanisole[128], a classic reagent for demonstrating O-dealkylase activity.

Gallo et al.[129, 129a] have also examined the cytochrome profiles of *C. tropicalis* grown on alkanes, and demonstrated the presence of a cytochrome P-450 and an NADPH-cytochrome c reductase. In a recent study by these authors, a new procedure was tested for the isolation of P-450-containing microsomes by lysis of protoplasts formed by the action of helicase on intact yeast cells[130]. This method was shown to be superior to other disruption techniques such as the French press or disintegration, but only if one considers P-450 levels in crude cell homogenates. After sucrose gradient centrifugation, the three procedures all gave comparable results. With the helicase treatment, no P-420 was detected, but could be formed by the action of protease on P-450. Attempts at reconversion were unsuccessful in this case. The microsomal P-450 isolated was remarkably stable, showing no spectral deterioration if stored at -20° for 15 days[130]. *Candida guilliermondii* also produces a form of P-450 (455 nm) catalyzing the oxidation of alkanes[130a].

For several years, *Saccharomyces cerevisae* has been under intense investigation because of the presence of an electron transport system similar to that in the endoplasmic reticulum of mammals[131]. The presence of a cytochrome P-450 in this organism grown under anaerobic conditions has been demonstrated by a number of groups[131–133, 133d, 133e], and also forms aerobically if cells are harvested prior to glucose depletion[133a, b]. This work was recently summarized by Wiseman[118]. The exact function of this enzyme is not fully delineated, although Wiseman[118] has suggested involvement in the ω-hydroxylation of oleic acid prior to its incorporation into endoplasmic reticulum or mitochondria[133], while others suggest involvement in yeast sterol synthesis[132]. There is a demonstrable reciprocal relationship between cytochrome P-450 levels and mitochondriogenesis, in that high glucose levels are necessary for suppressing mitochondrial formation and the stimulation of P-450 formation[131, 131a, 131b]. This interaction may also be linked by cyclic nucleotide formation[118].

The spectral and *catalytic stability* of P-450 from *S. cereviseae* was examined by Wiseman et al.[134], and the oxidized form appears to be more stable. Catalytic capability was followed by biphenyl-4-hydroxylase activity[135]; the enzyme also N-demethylates ethylmorphine, and aminopyrine[118]. The authors suggested[118] that additional stability might be attained by cross-linking or immobilization of the enzyme. Also, substrates exhibiting both type I and II binding have been found to protect the enzyme from denaturation by detergents[118].

Callen and Philpot[133b] utilized *S. cerevisiae* as a model system for the metabolic activation of a variety of promutagens. Conversion of two diploid strains was used as an indication of metabolic activities which was attributed to cytochrome P-450 dependent monooxygenase(s). The correlation of P-450 levels and conversion was impressive with nearly all of the promutagens tested.

Yoshida et al.[136, 136a] recently reported that a solubilized form of P-450 from *Saccharomyces cerevisiae* more closely resembles cytochrome P-448, in analogy to the 3-MC-induced cytochrome in mammals. These authors observed type II and modified type II binding, but no type I binding of substrates with this solubilized enzyme, which is capable of hydroxylating aniline and N-demethylating aminopyrine. In a second study[137], P-448 was obtained in 75% pure form using proteolytic procedures for isolation. The system was spectrally stable when stored at $-80\,^{\circ}$, and showed no contamination by cytochrome P-420. Studies also indicated it to be more stable to denaturing agents than mammalian cytochrome P-450, an important consideration in large scale isolation and utilization.

4.3 Bacteria

Perhaps the most intensively studied source of cytochrome P-450 is the fluorescent pseudomonad, *Pseudomonas putida.* As first described by Gunsalus et al.[138], this system is capable of conducting a stereospecific hydroxylation of camphor, forming the 5-*exo*-alcohol. Molecular oxygen was required, as well as NADH[138]. This has been found to be the case in nearly all bacterial P-450 systems; that is, a dependence on NADH rather than NADPH. Due to the very specific catalytic capability of this enzyme system, it was designated cytochrome P-450_{cam}. Research with this camphor hydroxylase has been quite extensive, and was recently reviewed by Gunsalus et al.[139], as well as Wiseman[118].

The P-450_{cam} system possesses three proteins that have been resolved. These individual components are: 1, an NADH coupled Putidaredoxin reductase; 2, an iron-sulphur protein termed Putidaredoxin; 3, a cytochrome P-450[140]. All components are necessary for hydroxylation to occur. Similar components were reported for an enzyme catalyzing the ω-hydroxylation of fatty acids by *Pseudomonas oleovorans,* although a P-450 does not appear to be involved[141]. The scheme proposed by Gunsalus[138] for P-450_{cam} is presented in Fig. 8.

Cytochrome P-450_{cam} has been purified[142] and crystallized[143] and consists of a single polypeptide, with a molecular weight of 44,000–46,000. Higher levels of enzyme were afforded by substrate protection, although this was not an absolute necessity. Any inactive form (P-420) generated during the isolation or by organic solvent denaturation, could be reactivated using sulfhydryl reagents, especially cysteine[144]. Glycerol and other

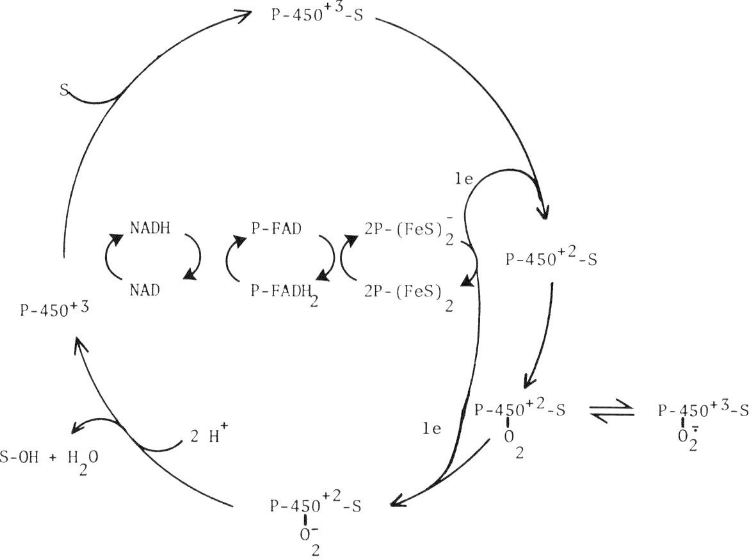

Fig. 8. Oxidation of substrate (S) *via* cytochrome P-450_{cam}
P-FAD = putidaredoxin reductase; P-(FeS) = putidaredoxin

polyols were unable to reverse it. These results suggest a modification of a sulfhydryl groups of the apoprotein.

Dus et al.[145] have compared the immunological cross-reactivity of P-450 from phenobarbital-induced rabbit liver microsomes and P-450_{cam}. A close relation was shown based on competitive binding and inhibition of catalytic activity. Treatment of the proteins with cyanogen bromide yielded heme-containing peptides, with amino acid compositions that were quite similar in the heme region[145].

Broadbent et al.[146] found that a *Nocardia* species (NH_1) grown on *iso*-vanillate, conducted O-dealkylation of *p*-alkylphenyl ethers *via* a P-450 monooxygenase, dubbed P-450_{npd} (npd = *Nocardia para* dealkylation). In a particularly elegant study, this cytochrome was obtained by polyacrylamide gel electrophoresis or ion exchange chromatography and shown to be homogenous by several criteria[146]. A molecular weight of 42,000–45,000 was calculated and the amino acid composition determined. Comparisons of P-450_{npd} and P-450_{cam} indicated similarities in isoelectric point, binding spectra, and molecular weight; major differences included markedly different substrate specificities and the lack of a high spin state for the iron present in P-450_{npd}[146].

Cardini and Jurtshuk[147] examined an alkane hydroxylase from *Corynebacterium species* (7EIC) and demonstrated that if functioned as a typical P-450-type monooxygenase. Both NADH and molecular oxygen were required for the conversion of octane to octanol in cell-free systems. Catalysis was sensitive to CO but not to cyanide, as would be expected, and a typical 450 nm peak was observed for the reduced P-450-CO complex. The presence of cytochrome P-420 was also demonstrated. The levels of cytochrome P-450 in this preparation were reported to be 0.25 nmoles mg^{-1} protein. A "particulate" form of cytochrome P-450, associated with lipids, was obtained from this culture[148]. The flavoprotein isolated, serves in the capacity to reduce cytochrome P-450 or cytochrome c, and requites NADH.

The solubilization and partial purification of a cytochrome P-450 from a nitrogen-fixing *Rhizobium bacteroids* was reported by Appleby[149]. The bacterial P-450 demonstrated a molecular weight of 50,000 and was converted to P-420 by prolonged incubation, in a reduced form, with CO. Evidence based on difference spectra were used to predict a schematic formulation of the catalytic center. Although the role of such an enzyme system was not shown, it was suggested that it may function as an azoreductase, in analogy to such, a function attributed to mammalian P-450[150].

Miura and Fulco[151, 152] have examind an ω-2 fatty acid hydroxylase from *Bacillus megaterium* (ATCC 14581), and have shown it to be a soluble enzyme (100,000 x *g* supernatant) and quite stable. The enzyme system required NADPH and molecular oxygen. Later studies demonstrated inhibition by CO, but not by azide or cyanide[153]. During fractionation on DEAE-cellulose, a component was removed which could be replaced by bacterial ferrodoxin from *Clostridium pasteurianum*[151]. Presumably the hydroxylase and reductase form a non-dissociating complex. Due to this interconvertability, this system was described as similar to P-450 systems from other bacteria, and, indeed, from mammalian adrenal cortex, in that all require an iron-sulfur protein for activity. This is in contrast to liver microsomal P-450 (and, apparently, fungal and yeast P-450) where no such dependence has been shown. The inhibition of this system by *p*-hydroxymercuribenzoate, a typical sulfhydryl modifying reagent, could be reversed by ferrodoxin, cysteine, or glutathione. It has been suggested that the mercurial interacts with a sulfur group or the iron-sulfur protein linking P-450 to the flavoprotein reductase[154].

Bacillus megaterium (ATCC 13368) has also been found to conduct the 15β-hydroxylation of progesterone, and a number of other 3-oxo-Δ^4-steroids[155]. This system requires NADPH, and is inhibited by CO, metyrapone, and SKF-525 A. A typical reduced cytochrome-CO complex spectrum was indicative of P-450. In analogy to mammalian P-450 hydroxylases, both sodium periodate and sodium chlorite can function as hydroxylating agents in this system, presumably by bypassing the reduction steps of the NADPH pathway (see Fig. 2). These authors have, in fact, suggested a new criteria for indicating P-450 involvement; support of hydroxylation by the noted oxidizing agents[155]. Also, because of the diverse number of microorganisms catalyzing various hydroxylations, usually with high regio- and stereospecificity, it was suggested that cytochrome P-450 from microbial species may prove invaluable in determining the nature of substrate specificity with such monooxygenases.

Berg et al.[156] have also fractionated the progesterone 15-β-hydroxylase of *B. megaterium* and found an NADPH dependent FMN containing flavoprotein called megaredoxin reductase, an iron-sulfur protein, megaredoxin, and cytochrome P-450_{meg}. Activity depends on all of these components. Thus, while similar to other bacterial P-450 systems based on a requirement for an iron-sulfur protein, *B. megaterium* appears unique because of its dependence on NADPH (with its reductase containing FMN) rather than NADH (with its reductase containing FAD).

5 Induction of Cytochrome P-450 Monooxygenases in Microorganisms

It should be expected that for monooxygenases of the P-450 type, as with other enzyme systems, that induction by substrates is a strong possibility. Indeed, many of the cases so far studied have shown induction of this type. Several of the species discussed above were, in fact, isolated by means of enrichment culture on substrate which *de facto* implies its metabolism. Fewer studies have been found where authors have examined to more classical P-450 inducers used routinely in mammalian work. In the limited number cases studied, parallels to the induction in mammalian species have been observed.

5.1 Fungi

The P-450 from the alkaloid producing strain of *Claviceps purpuria* examined by Ambike et al.[99] was shown to be induced two-fold by Pb. The increase in alkaloid production was shown to parallel P-450 induction. While enzyme levels were not dramatically induced with 3-MC, a typical spectral shift from 450 nm to 448 nm was observed[99]. Evidence based on protein synthesis inhibitors was given to indicate *de novo* P-448 synthesis with 3-MC. Tryptophan, used in the biosynthesis of the alkaloids, did not increase cytochrome P-450 levels, but was synergistic in this regard with Pb.

Using a benzo[a]pyrene hydroxylase assay, Ferris et al.[112] examined a number of inducers of the cytochrome P-450 in *Cunninghamella bainieri* (ATCC 9244). Surprisingly, benzene, 3-MC, Pb, and sodium barbital were all inactive in this respect. However, a 10-fold induction of hydroxylase activity was observed when phenanthrene was employed as the inducing agent. Unfortunately, these results were complicated by low cell yields[112]. Since aryl hydrocarbon hydroxylase activity is catalyzed by P-448 in mammalian species, it is not surprising that barbiturates demonstrated no effect. It is surprising, however, that 3-MC, a classic mammalian inducer of P-448 and aryl hydrocarbon hydroxylase did not appropriately function in the *C. bainieri*. It would be of further interest to examine the effects of these inducers on the other diverse P-450-catalyzed reactions in this organism[110].

Convincing evidence has been provided by Breskvar and Hudnik-Plevnik[157] that the progesterone hydroxylase of *Rhizopus nigricans*, a P-450 enzyme system, is induced by the substrate. Induction was shown in both growing cultures and in sucrose-buffer cell suspension, the latter showing higher induction. Inhibitor studies with cycloheximide revealed that increased hydroxylation was due to true induction, i.e. *de novo* enzyme synthesis.

Cerniglia and Gibson[121] examined the influence of classic inducers on the metabolism of naphthalene by their *Cunninghamella elegans* strain. Cultures were grown in the presence of the inducers. Naphthalene resulted in a five-fold increase in its own metabolism as compared to cultures grown in glucose. Indeed, the culture used in much of their work was isolated by enrichment culture based on its ability to utilize hydrocarbons such as naphthalene as a sole carbon sources Pb showed a three-fold increase and 3-MC a two to three-fold increase in naphthalene oxidation compared to glucose. It would be interesting to know if the 3-MC induced cytochrome represented P-448 rather than P-450, since there was no apparent drastic change in metabolic profile between Pb (a classic P-450 inducer) and 3-MC (a classic P-488 inducer)[121].

5.2 Yeasts

All indications are that the cytochrome P-450 of *Candida tropicalis* (LM-7) is induced by *n*-alkanes, with tetradecane being most widely used. This strain was, in fact, isolated based on its ability to use long chain *n*-alkanes as the major carbon source. Investigators often maintain this culture on agar slants coated with a tetradecane film[128]. Lebeault et al.[127] showed that the hydroxylase activity, the P-450 levels, and NADPH-dependent cytochrome c reductase were all induced by the presence of tetradecane dispersed in the reaction under high aeration and agitation. Glucose does not induce P-450 in this culture.

Studies by Gallo et al.[129, 130] indicated similar results. In fact, it was reported that P-450 is absent if the culture is grown on classical substrates such as glucose or acetate. Of further importance is that hydrocarbons not only induce those enzymes concerned with hydroxylation, but also certain alcohol and aldehyde dehydrogenases. A number of other heminic compounds were induced by hydrocarbons in this organism, but the physiological significance of these substances is not yet known[129, 130]. The fatty acid metabolizing system of *Candida intermedia* (NRRL-Y6328-1) also shows induction by growth on hydrocarbons, which was the basis of the original isolation[127a].

Wiseman and Lim[158] found that the cytochrome P-450 of *Saccharomyces cereviseae* may be induced by Pb in the growth medium containing 0.5% glucose. It should be remembered from an earlier discussion that this is the reverse situation concerning glucose vs. P-450 levels; i.e. high levels (20%) of glucose are necessary for mitochondrial repression and production of sizeable quantities of P-450. Normally, at 0.5% glucose, no P-450 is formed, however, the Wiseman and Lim[158] study indicates that the lower carbohydrate levels are conducive to P-450 production if Pb is present and, indeed, resultant cytochrome levels approach those seen in 20% glucose. No induction above normal levels was seen in cultures grown in 20% glucose with Pb. Deadaptation and inhibitor studies indicated *de novo* enzyme synthesis which is associated with true induction[158].

5.3 Bacteria

Broadbent et al.[146] describe a *Nocardia sp.* that possesses a P-450-isovanillate-inducible O-dealkylase with activity towards a series of *p*-alkylphenyl ethers. The P-450_{cam} of the fluorescent *P. putida* (Pp61) is probably induced by the substrate, camphor, since the strain was first isolated by the enrichment culture technique using camphor as the sole carbon source[138]. Although apparently not examined, it is highly unlikely that this system would be induced by classical inducers such as Pb. The narrow substrate specificity of this system would probably not allow for metabolism of the inducer which is one (though not the exclusive) requisite for induction[66].

A *Corynebacterium sp.* (7ElC) was also grown on a sole carbon source of *n*-octane, usually applied in the vapor state as hydrocarbon-saturated air[148]. Careful examination of cultures grown on another carbon source not involved in the metabolism of octane, namely acetate, indicated a six-fold reduction of the cytochrome which was also only apparent as its P-420 form. The induction was shown both spectrally and catalytically. Fractionation of the hydroxylating system indicated that the flavoprotein component is absent in the non-induced (acetate grown) cells[148].

Edelson and McMullen[159] recently examined the O-demethylase activity of *E. coli*, a common constituent of gut microflora. Studies involving the dealkylation of *p*-nitroanisole indicated that such activity could be induced by Pb to an activity 1.5 times that of uninduced cells. The inducer only moderately reduced cell-growth. No stimulation was observed by the addition of exogenous cofactors, but inhibition by CO was apparent. This study is exciting because of the numerous implications of gut microflora in the metabolism of drugs and other xenobiotics[160, 161]. The possibility that agents administered to mammals may influence the xenobiotic metabolizing activity of intestinal microorganisms may open a new and potentially fruitful area of research.

5.4 Effects of Surfactants and Antioxidants

Because microbial transformation studies are frequently carried out in the presence of additives such as surfactants and antioxidants, it is important to understand influences of these types of agents on P-450 monooxygenases. Surfactants are commonly employed in microbial work as wetting agents or as antifoams. When used as the latter, it is generally considered that they replace normal surface-active agents of the medium or culture that cause foaming, but cannot support foam-formation themselves[162]. In addition, it can be expected that dispersion of water insoluble compounds by enhanced wetting, decreased flotation, or micellular solubilization may be beneficial[8]. For example, metabolism of progesterone wetted with 0.01% Tween 80 was increased more than 2-fold with *Aspergillus ochraceus*[163]. Whitworth et al.[164] found marked growth enhancement of *Candida lipolytica* (ATCC 8661) on dodecane when surfactants such as Tweens and Spans were used. This was probably a result of enhanced dispersion of the hydrocarbon, as well as the ability of the surfactants to support growth.

The various types of agents available and their sources have been described elsewhere[162, 165], as were the mechanism of *antifoaming action*[166]. Basically, the criteria for a good antifoam are, activity at low concentrations, no deleterious interaction with the microorganism (or the desired metabolic process), and no interference with the assay or isolation of products[162]. Most problems are encountere in stirred vessels where antifoams are more commonly employed. In such cases, Bryant[165] suggests that the antifoam may affect oxygen transfer at any of three critical steps, the transfer of air across the gas/liquid interface, transfer of gases in the fluid medium, or transfer of dissolved gases to the surface of the microorganism which may be rate-limiting.

These mechanisms suggest that one should not blindly add antifoams to reactions unless the effects of such agents on biochemical properties being examined are evaluated. One might expect this to be especially true of oxidative reactions such as those involving P-450 systems. The role of dissolved oxygen in controlling a number of microbiological systems has been discussed[167).

Authors generally agree that choosing an antifoam is empirical, but it has been suggested that silicones appear to provide good antifoam action with bacteria, while polyglycols are more effective with fungi[162)]. Some investigators prefer "natural surfactants" such as vegetable oils, since they may be used as a carbon source and hence, do not tend to accumulate; they also act as slow-release carbon stores[168)]. In any case, the amount of antifoam used should be minimal, and may be reduced further by using a synergistic "carrier"[162)].

In one of the studies by Duppel et al.[128)], the influence of Antifoam B on *Candida tropicalis* was examined. Since maximal growth depended on high stirring rates for increased aeration and tetradecane dispersion, severe foaming resulted which could not be prevented by the antifoam. The additional important observation was made that, when using Antifoam B, P-420 was formed at the expense of P-450! Berg et al.[156)], used Antifoam B in the submers production of *Bacillus megaterium* for a P-450 determination. Apparently, no difficulties were encountered in this case.

Wiseman et al.[133, 133b)] examined the influence of Tween 80, an oleic acid derivative, on P-450 production in *Saccharomyces cerevisiae.* Tween 80 was required, along with ergosterol, for the biogenesis of mitochondria induced by aerobic conditions or glucose limitation. The surfactant was shown to double P-450 levels after 50 hours using a 3% glucose medium. It was proposed that this was due to decreased cytochrome degradation, rather than induction of the enzyme system.

A number of cases can be cited in which antioxidants or other stabilizing agents have been utilized to prevent the decomposition of cytochromes P-450 from microbial sources. Indeed, several cases will be noted below in which the denatured form, P-420, was reverted to P-450 in the presence of antioxidants. In such cases, reconversion is often based on spectral reconversion, which is only presumptive of reconverted catalytic activity. Numerous agents have been employed including polyols and sulfhydryl compounds which have also been shown to allow reconversion of mammalian P-420[101)]. It was proposed that such agents uncouple the denaturing detergents or sulfhydryl compounds which cause the formation of P-420[101)].

A cytochrome P-450 from *Claviceps purpurea* was converted to P-420 with sodium cholate[100)]. This was reversed to varying degrees with glycerol; 10% (V/V) solutions providing nearly quantitative reconversion. Steapsin treatment (to remove b_5) also caused the P-450 to P-420 conversion, which could be prevented by glycerol.

Murphy et al.[119)] found that the *m*-cresol hydroxylase activity of *Penicillium patulum* was most stable in cell-free form if 10% glycerol was incorporated into the buffer. The vast majority of authors who have examined P-450 from microorganisms have utilized a combination of stabilizing agents in cell-free systems including dithiothreitol, glycerol and/or EDTA[114, 121, 125, 128, 129, 136, 137, 143)]. In addition, some of these agents have been shown to reverse the P-450 to P-420 process. For example, the P-420_{cam} from *P. putida* could be converted to P-450_{cam} with sulfydryl compounds such as cysteine, but not with glycerol or other polyols[144]. The ω-2 hydroxylase of *B. megaterium* was inactivated with *p*-hydroxymercuribenzoate, which could be reversed with cysteine or glutathione[153]. The P-450 from this culture catalyzes the 15β-hydroxylation of steroids and is somewhat denatured during purification[156]. The P-420 formed was converted to P-450 using sodium borohydride[156].

Other investigators have encountered difficulty in affecting the conversion of P-420 to P-450, or in stabilizing P-450. For example, the P-420 formed during microsomal isolation from *C. tropicalis* was not reversed by using reducing agents or polyols[130]. Stabilization was afforded using bovine serum albumin and EDTA during fractionation. It has been suggested by Scarpa and Lindsay[169] that such agents protect microsomal membranes and prevent denaturation of cytochromes by the inhibition of endogenase proteases and by inhibition of phospholipids which are dependent on free calcium ions.

Ferris et al.[111] found that *p*-chloromercuribenzoate inhibited the aryl hydrocarbon hydroxylase activity of *C. bainieri*, presumably by conversion to P-420 or by binding to an associated reductase. Hydroxylase activity in the cell-free system decreased on storage at 5°, and this could not be prevented by using sodium borohydride, mercaptosuccinate, dithiothreitol, or a number of other agents[111].

Some workers[109] have stabilized the P-450 from *Torulopsis sp.* with glyceral during isolation, but found no stabilizing effect attributable to sulfhydryl compounds or ascorbate. Lebeault et al.[127] found that the P-450 from *C. tropicalis* was not readily autooxidizable, but could be stabilized with glycerol or dithiothreitol. Duppel et al.[128] found that P-420 formation during the fractionation of the P-450 hydroxylase from *C. tropicalis* could not be suppressed using glycerol. However, the solubilized form of P-450 from this organism is the presence of glycerol and dithiothreitol showed no P-420[128].

It is obvious from such cases that the choice of the appropriate antioxidant(s) for a particulat microbial system involving P-450 is empirical. It should also be apparent, however, that specificity may be exhibited in such systems, and that a variety of agents may have to be examined before an appropriate agent is found. Such studies are not only important from a biochemical standpoint, but also for the practical application of such systems to larger, preparative-type applications.

6 Case of Biphenyl

Biphenyl has become a model substrate for studying P-450-catalyzed hydroxylations. Two reasons are that its two principal metabolites, 2-hydroxybiphenyl and 4-hydroxybiphenyl, are relatively stable chemically and are quite readily analyzed [171–174]. Additionally, and more importantly, the two hydroxylated products appear to be formed to differing extents by distinct forms of cytochrome P-450. Thus, their relative production can serve as an indirect marker for induction. Indeed, the reported differential induction of biphenyl-2-hydroxylase activity by carcinogens in certain mammals[175–182] has intrigued a number of workers who propose[177–182] that such a system may serve as the basis of a screen for carcinogenic compounds. Recent investigations in a number of laboratories[134, 186] suggest that hydroxylations parallel to those affected by mammalian systems also occur in certain eurkaryotic microorganisms.

6.1 Hydroxylations by Mammalian vs. Microbial Systems

As indicated in Fig. 9, biphenyl is converted to a variety of hydroxylated metabolites *in vivo*. The proportions of phenolic metabolites varies with species, however, the 4-hydroxylated derivatives (principally 4-hydroxybiphenyl) are most important quantitatively[187–190]. The *in vitro* hydroxylation of biphenyl by liver microsomal preparations has also been extensively studied[175–182, 191]. Practically all of the phenolic metabolites found in Fig. 9 have been detected *in vitro*. However, the greatest amount of effort has been placed on the formations of 2-, 3-, and 4-hydroxybiphenyls (see below) which are quite readily separable by gas chromatography (GC) or high performance liquid chromatography (HPLC)[174, 182, 183].

There is considerable species and age differences observed in the production of 2- and 4-hydroxybiphenyls. Biphenyl 4-hydroxylase activity is observed in liver microsomal preparations from all of the greater than eight mammalian species investigated[171, 191]. In contrast, relatively few animals possess 2-hydroxylase activity[171, 191]. Interestingly, formation of 2-hydroxybiphenyl is more prevalent in microsomal preparations from younger vs. older animals[171]. Earlier studies suggested that the different biphenyl hydroxylases possessed varying kinetic properties[171]. It is now widely believed (from studies with inhibitors and inducing agents; see below) that the 2-hydroxylase activity is principally due to cytochrome P-448. In contrast, cytochrome P-450 is primarily responsible for the 4-hydroxylation of biphenyl[181, 182, 185].

A recent article by Billings and McMahon[182] has shed light on the mechanism of formation of 2-, 3-, and 4-hydroxybiphenyl by mammalian liver microsomal monooxygenases. Their studies support intermediary arene oxides in the production of 2- and 4-hydroxybiphenyl (see Fig. 10). In contrast, the 3-hydroxy-isomer is probably formed by a direct insertion mechanism as indicated by a significant deuterium isotope effect (K_H/K_D=1.27-1.45) observed during the 3-hydroxylation of biphenyl-d_{10}. Isotope effects less than 1.2 (K_H/K_D) were observed during the formation of 2- and 4-hydroxybiphenyl by rat liver microsomes thereby supporting the arene oxide mechanism in the generation of these metabolites. Interestingly, Billings, and McMahon[182] have accumulated evidence suggesting that 3,4-dihydroxybiphenyl is formed from the parent hydrocarbon by successive monohydroxylations rather than dehydrogenation of a dihydrodiol. These results may be of general importance in catechol formation and should be explored further.

Smith and Rosazza[5] have studied the metabolism of biphenyl by eleven microorganisms (nine fungi and two *Actinomycetes*) that were chosen based on their reported ability to metabolize aromatic substrates. In analogy with mammalian systems,

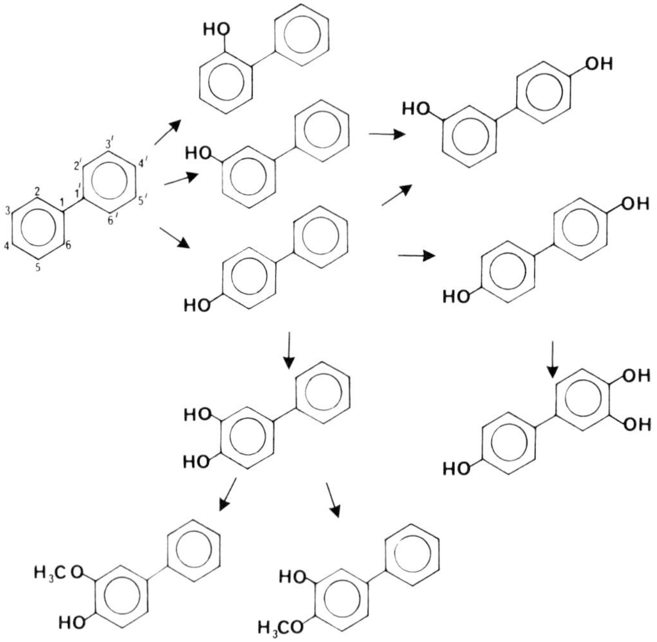

Fig. 9. Hydroxylated metabolites of biphenyl formed *in vivo* in various mammals (see Refs.[187–190])

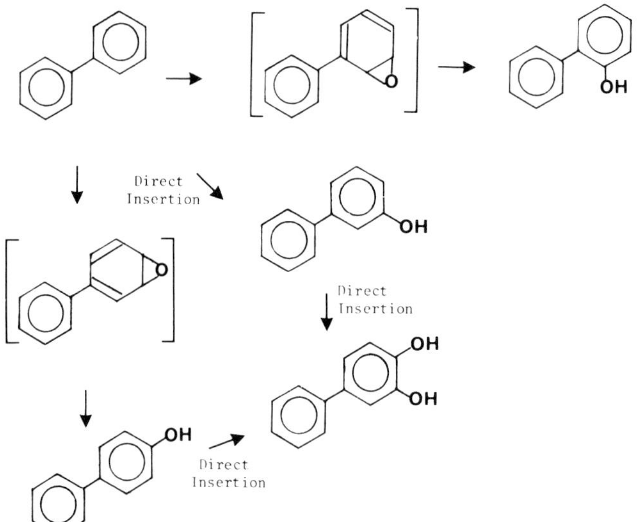

Fig. 10. Hydroxylations of biphenyl by liver monooxygenases (see Refs.[175–182, 191])

4-hydroxylation predominated in the majority of cultures that metabolized biphenyl; 4-hydroxybiphenyl was produced by six cultures while 4,4'-dihydroxybiphenyl was detected with two organisms. Two microorganisms produced both 2-hydroxybiphenyl and 4-hydroxybiphenyl with the former metabolite predominating in a *Helicostylum piriforme* and the 4-hydroxy isomer predominating in an *Aspergillus niger*[5]. A *Streptomyces rimosus* produced 2-hydroxybiphenyl exclusively, albeit in low yield[5]. We have generally confirmed these results recently in our laboratories through use of a sensitive and selective GC assay[174], and we are pursuing the identification of other trace metabolites from the noted organisms. It is interesting that Gibson's group[192] have recently noted that four organisms out of thirteen fungi and one yeast examined, produced 2-hydroxybiphenyl, 4-hydroxybiphenyl, and 4,4'-dihydroxybiphenyl. Furthermore, three of these cultures produced 2,2'-dihydroxybiphenyl[192]. Wiseman et al.[133, 134] showed that a P-450-containing microsomal fraction from a *Saccharomyces cerevisiae* is capable of catalyzing the 4-hydroxylation of biphenyl.

A number of investigators have isolated bacterial strains capable of utilizing biphenyl as a sole carbon source, including *Achromobacter*[193, 194], *Pseudomonas*[195, 196], *Beijerinckia*[197] and several unidentified species[198, 199]. Based on the results of these studies, however, it appears that all of these activities are probably attributable to diooxygenase rather than monooxygenase enzymes.

6.2 Induction of 2- vs. 4-Hydroxylase Activities

Induction of mice and rats with 3-MC and other compounds that increase cytochrome P-448, causes preferential increases in 2-hydroxylation of biphenyl[175, 176, 182]. In contrast, Pb induces principally 4-hydroxylase activity which is attributable to increased *de novo* synthesis of cytochrome P-450[181, 182]. Analogous, though somewhat less pronounced effects, are observed in hamsters; possibly due to the constitutive 2-hydroxylase in this species[191].

Recently, McPherson et al.[181] reported that preincubation of rat liver microsomes with carcinogenic compounds such as 3,4-benzo[a]pyrene, 3-MC, and safrole selectively induced biphenyl-2-hydroxylase activity. At the same time, 4-hydroxylase activity was inhibited or not effected at all. It was proposed that the preferential induction of 2-hydroxylase *in vitro* might serve as the basis for a screening procedure for carcinogens[181]. However, Billings, and McMahon[182] report an inability to reproduce the findings of McPherson et al.[181] using rat and mice liver microsomes. The former authors[182] did, indeed, observe an increase in biphenyl-2-hydroxylase activity by preincubation with the P-448 inducer, α-naphthoflavone (7,8-benzoflavone). However, a significant inhibition of 2-hydroxylase activity was observed following preincubation with benzo[a]pyrene and 3-MC[182]. This discrepancy may be related, in part, to spectral interferences arising from the metabolism of inducers as described by Burke et al.[183].

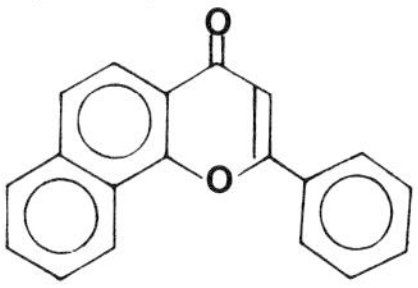

α-Naphthoflavone
7,8-benzoflavone

Since Billings and McMahon[182] utilized a GC assay, their conclusions should be more valid. However, the induction of biphenyl-hydroxylase activity deserves further study. Of particular importance is whether or not the *in vitro* induction of biphenyl-2-hydroxylase (P-448) is real.

Such a study should also address the problem of P-450-P-448 interconvertability and the importance of individual forms of P-450 in the various biphenyl hydroxylations.

Utilizing fungal species chosen earlier for their ability to metabolize biphenyl[5] and a sensitive and selective GC assay[174], we have initiated studies to examine whether the differential induction of hydroxylase activities seen with mammalian species also occurs in microorganisms. The cultures being studied include *Helicostylum piriforme* (QM-6945) for the production of 2-hydroxybiphenyl, *Cunninghamella echinulata* (ATCC 9244) for the production of 4-hydroxybiphenyl, and *Cunninghamella elegans* (ATCC 9244) for the production of 2- and 4-hydroxyisomers. Classic carcinogenic and non-carcinogenic inducing agents are being examined, as well as several polychlorinated biphenyls based on results with mammalian systems[85–87]. Having studied the effect of a number of microbial growth parameters on the qualitative and quantitative profiles of hydroxybiphenyl metabolites[200], inducer studies have commenced[201]. It is our hope that results from these investigations will aid in delineating the mechanism(s) of 2- and 4-biphenyl hydroxylase induction in microorganisms.

7 Outlook

7.1 Induction of Mammalian P-450 Monooxygenases

Because of its basic biochemical and toxicological importance, the induction of mammalian monooxygenases will receive even greater attention in the future. Inducing agents have been advantageously employed to enrich certain liver microsomal P-450 monooxygenases which allowed subsequent isolation and purification. It seems likely that additional inducing agents will be discovered that may permit isolation of the remaining uncharacterized P-450 monooxygenases that occur in mammals.

7.2 Induction of Microbial Monooxygenases

Sufficient evidence now exists for the widespread distribution of cytochromes P-450 in a significant number of diverse microorganisms, and the number discovered will undoubtedly increase. Such observations are of interest from a purely biochemical and teleological standpoint, i.e., not only in understanding the function(s) of such enzymes in lower forms of life, but also for comparative biochemical studies of metabolic capability between microorganisms and mammals. It is true that a number of P-450 systems thus far isolated, catalyze the conversion of a narrow range of substrates ($P\text{-}450_{cam}$[138], $P\text{-}450_{npd}$[146]). Nonetheless, other microbial P-450 systems (particularly eukaryotic, such as that from *C. bainieri*[111,112] and *C. tropicalis*[127, 128] catalyze a range of metabolic reactions rivalling those of mammalian liver.

As interest grows in the large scale preparations of cytochrome P-450 and in the use of biological systems to conduct specific reactions, study of microbial P-450 will continue to flourish. It appears that a major emphasis needs to be placed on evaluating conversions of P-420 to P-450 and *vice versa* not only by spectral analysis, but also by catalytic evaluation, since this represents the ultimate criterion for utilizing microbial cytochrome P-450's.

8 Acknowledgements

This work was supported in part by grant F-690 from the Robert A. Welch Foundation. The authors are grateful to Terry T. Sayther for performing some preliminary literature work that contributed to this paper.

9 References

1. Miller, E.C., Miller, J.A., Brown, R.P.: Cancer Res. *12*, 282 (1952)
2. Conney, A.H., Miller, E.C., Miller, J.A.: Cancer Res. *16*, 450 (1956)
3. Breckenridge, A.: In: Enzyme induction. Parke, D.V. (ed.), p. 273. New York: Plenum Press 1975
4. Conney, A.H.: Pharmacol. Rev. *19*, 317 (1966)
5. Smith, R.V., Rosazza, J.P.: Arch. Biochem. Biophys. *161*, 551 (1974)
6. Smith, R.V., Rosazza, J.P.: J. Pharm. Sci. *64*, 1737 (1975)
7. Smith, R.V., Rosazza, J.P.: Biotechnol. Bioeng. *17*, 785 (1975)
8. Smith, R.V., Acosta, D., Rosazza, J.P.: Adv. Biochem. Eng. *5*, 69 (1977)
9. Ullrich, V., Kremers, P.: Arch. Toxicol. *39*, 41 (1977)
10. Coon, M.J. et al.: In: Drug metabolism concepts. Jerina, D.M. (ed.), p. 46. Washington, D.C.: Amer. Chem. Soc. 1977
11. Lu, A.Y.H., West, S.B.: Pharmacol. Ther. A *2*, 337 (1978)
12. Watanuki, M., Tilley, B.E., Hall, P.F.: Biochemistry *17*, 127 (1978)
13. Pedersen, J.I.: FEBS Lett. *85*, 35 (1978)
14. Fahl, W.E., Jefcoate, C.R., Kasper, C.B.: J. Biol. Chem. *253*, 3106 (1978)
15. Acosta, D., Anuforo, D., Smith, R.V.: In Vitro *14*, 428 (1978)
16. Gibaldi, M., Perrier, D.: In: Drug metabolism reviews. Di Carlo, F.J. (ed.), Vol. 3, p. 185. New York: Marcel Dekker 1974–1975
17. Fang, W.-F., Strobel, H.W.: Arch. Biochem. Biophys. *186*, 128 (1978)
18. Kominami, S., Mori, S., Takemori, S.: FEBS Lett. *89*, 215 (1978)
19. Mazel, P.: In: Fundamentals of drug metabolism and distribution. LaDu, B.N., Mandel, H.G., Way, E.L. (eds.), p. 527. Baltimore: William & Wilkins 1971
20. Mailman, R.B. et al.: Gen. Pharmacol. *8*, 275 (1977)
21. Lu, A.Y.H., Coon, M.J.: J. Biol. Chem. *243*, 1331 (1968)
22. Lu, A.Y.H., Strobel, H.W., Coon, M.J.: Biochem. Biophys. Res. Commun. *36*, 545 (1969)
23. Lu, A.Y.H., Strobel, H.W., Coon, M.J.: Mol. Pharmacol. *6*, 213 (1970)

24. Lu, A.Y.H. et al.: J. Biol. Chem. *247*, 1727 (1972)
25. Levin, W. et al.: J. Biol. Chem. *249*, 1747 (1974)
26. van der Hoeven, T.A., Coon, M.J.: J. Biol. Chem. *249*, 6302 (1974)
27. Kamataki, T. et al.: Drug Metab. Disp. *4*, 180 (1976)
28. Philpot, R.M., Arnic, E.: Mol. Pharmacol. *12*, 483 (1976)
29. Yasukochi, Y., Masters, B.S.S.: J. Biol. Chem. *251*, 5337 (1976)
30. Imai, Y.: J. Biochem. *80*, 267 (1976)
31. Coon, M.J. et al.: Croat. Chem. Acta *49*, 163 (1977)
32. Bell, D.Y., Hodgson, E.: Gen. Pharmacol. *8*, 113 (1977)
32a. Gibson, G.G., Schenckman, J.B.: J. Biol. Chem. *253*, 5957 (1978)
33. Estabrook, R.W. et al.: Biochem. Biophys. Res. Commun. *42*, 132 (1971)
34. Bjorkhem, I.: Pharmacol. Ther. A *1*, 327 (1977)
35. Lu, A.Y.H., Levin, W., Kuntzman, R.: Biochem. Biophys. Res. Commun. *60*, 266 (1974)
36. Conney, A.H. et al.: Cancer Res. *17*, 628 (1957)
37. Wilsson, A., Johnson, B.C.: Arch. Biochem. Biophys. *101*, 494 (1963)
38. Gourlay, G.K., Stock, B.H.: Biochem. Pharmacol. *27*, 969 (1978)
39. Estabrook, R.W., Werringloer, J.: In: Drug metabolism concepts. Jerina, D.M. (ed.), p. 1. Washington, D.C.: Amer. Chem. Soc. 1977
40. Kadlubar, F.F., Morton, K.C., Ziegler, D.M.: Biochem. Biophys. Res. Commun. *54*, 1255 (1973)
41. Rahimtula, A.D., O'Brian, P.J.: Biochem. Biophys. Res. Commun. *60*, 440 (1974)
42. Ellin, A., Orrenius, S.: FEBS Lett. *50*, 378 (1975)
43. Rahimtula, A.D., O'Brian, P.J.: Biochem. Biophys. Res. Commun. *62*, 268 (1975)
44. Hrycay, E.G. et al.: FEBS Lett. *56*, 161 (1975)
45. Hrycay, E.G. et al.: Eur. J. Biochem. *61*, 43 (1976)
46. Nordblom, G.D., White, R.E., Coon, M.J.: Arch. Biochem. Biophys. *175*, 524 (1976)
46a. Jones, D.P. et al.: J. Biol. Chem. *235*, 6031 (1978)
47. Testa, B., Jenner, P.: Drug metabolism: chemical and biochemical aspects, p. 329. New York: Marcel Dekker 1976
48. Imai, Y., Sato, R.: J. Biochem. *62*, 464 (1967)
49. Franklin, M.R.: Xenobiotica *1*, 581 (1971)
50. Omura, T., Sato, R.: J. Biol. Chem. *239*, 2370 (1964)
51. McMahon, R.E. et al.: J. Med. Chem. *12*, 207 (1969)
51a. Netter, K.J., Jenner, S., Kajuschke, K.: Naunyn-Schmiedeberg's Arch. Pharmak. Exp. Path. *259*, 1 (1967)
51b. Netter, K.J., Kahl, G.F., Magnussen, M.P.: Naunyn-Schmiedeberg's Arch. Pharmak. Exp. Path. *265*, 205 (1969)
51c. Netter, K.J.: Proc. First Internat. Congr. Pharmacology, Stockholm 1961. Vol. VI, p. 213. New York: Pergamon 1962
52. Gillette, J.R., Brodie, B.B., LaDu, B.N.: J. Pharmacol. Exp. Ther. *119*, 532 (1957)
53. Orrenius, S.: J. Cell. Biol. *26*, 713 (1965)
54. Oshino, N., Imai, Y., Sato, R.: Biochim. Biophys. Acta *128*, 13 (1966)
55. Kitada, M. et al.: Jap. J. Pharmacol. *27*, 601 (1977)
56. Matsubara, T., Rouchi, A.: Jap. J. Pharmacol. *27*, 701 (1977)
57. Selander, H.G., Jerina, D.M., Daly, J.W.: Arch. Biochem. Biophys. *164*, 241 (1974)
58. Holder, J. et al.: Proc. Nat. Acad. Sci. U.S.A. *71*, 4356 (1974)
59. Selander, H.G., Jerina, D.M., Daly, J.W.: Arch. Biochem. Biophys. *168*, 309 (1975)
60. Nerland, D.E., Mannering, G.J.: Drug Disp. Metab. *6*, 150 (1978)
61. Burke, M.D., Mayer, R.T.: Drug Metab. Disp. *2*, 583 (1974)
62. Burke, M.D., Prough, R.A., Mayer, R.T.: Drug Metab. Disp. *5*, 1 (1977)
62a. Netter, K.J., Seidel, G.: J. Pharmacol. Exp. Ther. *146*, 61 (1964)
63. Nakatasugawa, T., Dahm, P.A.: Biochem. Pharmacol. *16*, 25 (1967)
64. Lotlikar, P.D., Zaleski, K.: Biochem. J. *150*, 561 (1975)

65. Schulte-Hermann, R.: CRC Crit. Rev. Toxicol. *2*, 97 (1974)
65a. Bock, K.W., Remmer, H.: In: Heme and hemoproteins, handbook of experimental pharmacology. DeMatteis, F., Aldridge, W.N. (eds.), Vol. 44, p. 49. Berlin: Springer 1978
66. Parke, D.V.: In: Enzyme induction. Parke, D.V.. (ed.), p. 207. New York: Plenum Press 1975
67. Gillette, J.R.: Metabolism *20*, 215 (1971)
68. Gillette, J.R.: Ann. N.Y. Acad. Sci. *179*, 43 (1971)
69. Haugen, D.A., Vander Hoeven, T.A., Coon, M.J.: J. Biol. Chem. *250*, 3567 (1975)
70. Haugen, D.A., Coon, M.J., Nebert, D.W.: J. Biol. Chem. *251*, 1817 (1976)
71. Welton, A.F., Aust, S.D.: Biochem. Biophys. Res. Commun. *56*, 898 (1974)
72. Ryan, D. et al.: J. Biol. Chem. *250*, 3567 (1975)
73. Atlas, S.A., Nebert, D.W.: Arch. Biochem. Biophys. *175*, 495 (1976)
74. Chhabra, R.S. et al.: Chem. Biol. Interact. *15*, 21 (1976)
75. Lu, A.Y.H. et al.: Ann. N.Y. Acad. Sci. *212*, 156 (1973)
76. Kumaki, K. et al.: J. Biol. Chem. *253*, 1048 (1978)
77. George, W.J., Tephly, T.R.: Mol. Pharmacol. *4*, 502 (1968)
78. Thompson, J.A., Holtzman, J.L.: Drug Metab. Disp. *5*, 9 (1977)
79. Nerland, D.E., Mannering, G.J.: Drug Metab. Disp. *6*, 150 (1978)
80. Poland, A., Glover, E.: Mol. Pharmacol. *10*, 349 (1974)
81. Nebert, D.W. et al.: J. Cell Physiol. *85*, 393 (1975)
82. Alvares, A.P., Bickers, D.R., Kappas, A.: Proc. Nat. Acad. Sci. U.S.A. *70*, 1321 (1973)
83. Alvares, A.P., Kappas, A.: J. Biol. Chem. *252*, 6373 (1977)
84. Ryan, D.E., Thomas, P.E., Levin, W.: Mol. Pharmacol. *13*, 521 (1977)
85. Poland, A., Glover, E.: Mol. Pharmacol. *13*, 924 (1977)
86. Poland, A., Glover, E., Kende, A.S.: J. Biol. Chem. *251*, 4936 (1976)
87. Poland, A. et al.: Science *194*, 627 (1976)
87a. Trager, W.F.: In: Drug metabolism concepts. Jerina, D.M. (ed.), p. 81. Washington, D.C.: Amer. Chem. Soc. 1977
88. Wattenberg, L.W.: Toxicol. Appl. Pharmacol. *19*, 54 (1971)
89. Vesel, E.S.: Science *157*, 1057 (1967)
90. Jori, A. et al.: Eur. J. Pharmacol. *9*, 362 (1970)
91. Parke, D.V., Rahman, H.: Biochem. J. *113*, 12 P (1969)
92. Parke, D.V., Rahman, K.Q., Walker, R.: Biochem. Soc. Trans. *2*, 615 (1974)
93. Bresnick, E.: Pharmacol. Ther. A. *2*, 319 (1978)
94. Khandwala, A.S., Kasper, C.B.: Biochem. Biophys. Res. Commun. *54*, 1241 (1973)
95. Watanabe, M. et al.: Gann *66*, 399 (1975)
96. Rogan, E.G., Cavalieri, E.: Biochem. Biophys. Res. Commun. *58*, 1119 (1974)
97. Rogan, E.G., Mailander, P., Cavalieri, E.: Proc. Nat. Acad. Sci. U.S.A. *73*, 547 (1976)
98. Bresnick, E. et al.: Arch. Biochem. Biophys. *181*, 257 (1977)
99. Ambike, S.H., Baxter, R.M., Zahid, N.D.: Phytochem. *9*, 1953 (1970)
100. Ambike, S.H., Baxter, R.M.: Phytochem. *9*, 1959 (1970)
101. Ichikawa, Y., Yamaro, T.: Biochim. Biophys. Acta: *131*, 490 (1967)
102. Geissman, T.A., Crout, D.H.G.: Organic Chemistry of Secondary Plant Metabolism, p. 561. San Francisco: Cooper 1969
103. Auret, B.J. et al.: Chem. Commun. 1585 (1971)
104. Daly, J.W., Jerina, D.M., Witkop, B.: Experientia *28*, 1129 (1972)
105. Jerina, D.M.: Chem. Tech. *1*, 120 (1973)
106. Boyd, D.R. et al.: J. Chem. Soc. Perkin I, 2438 (1976)
107. Fonken, G.S., Johnson, R.S.: Chemical oxidations with microorganisms. New York: Marcel Dekker 1972
108. Beukers, R., Marx, A.F., Zuidwig, M.H.J.: In: Drug design. Ariens, E.J. (ed.), Vol. 3, New York: Academic Press 1972

109. Sih, C.J., Rosazza, J.P.: In: Applications of biochemical systems in organic chemistry. Jones, J.B. (ed.), Part 1, p. 69. New York: Wiley 1976
110. Kieslich, K.: Microbial transformations of non-steroidal cyclic compounds. New York: Wiley 1976
111. Ferris, J.P. et al.: Arch. Biochem. Biophys. *156*, 97 (1973)
112. Ferris, J.P. et al.: Arch. Biochem. Biophys. *175*, 443 (1976)
113. Jones, E.R.H.: Pure Appl. Chem. *45*, 39 (1973)
114. Breskvar, K., Hudnik-Plevnik, T.: Biochim. Biophys. Res. Commun. *74*, 1192 (1977)
115. Breskvar, K., Hudnik-Plevnik, T.: Croat. Chem. Acta *49*, 207 (1977)
116. Schenkman, J.B., Remmer, H., Estabrook, R.W.: Mol. Pharmacol. *3*, 113 (1967)
117. Smith, R.V. et al.: Appl. Environ. Microbiol. *31*, 448 (1976)
118. Wiseman, A.: In: Topics in enzyme and fermentation biotechnology. Wiseman, A. (ed.), Vol. 1, p. 172. London: Halsted Press 1977
119. Murphy, G. et al.: Eur. J. Biochem. *49*, 443 (1974)
120. Cerniglia, C.E., Gibson, D.T.: Appl. Environ. Microbiol. *34*, 363 (1977)
121. Cerniglia, C.E., Gibson, D.T.: Arch. Biochem. Biophys. *185*, 121 (1978)
122. Schäfer-Ridder, M., Brocker, U.: Angew. Chem. *15*, 228 (1976)
123. Chen, C., Tu, M.: Biochem. J. *160*, 805 (1976)
124. Herbes, S.E., Schwall, L.R., Williams, G.A.: Appl. Environ. Microbiol. *34*, 244 (1977)
125. Davies, J.S., Wellman, A.M., Zajic, J.E.: Appl. Environ. Microbiol. *32*, 14 (1976)
126. Heinz, E., Tulloch, A.P., Spencer, J.F.T.: Biochim. Biophys. Acta *202*, 49 (1970)
127. Lebeault, J.M., Lode, E.T., Coon, M.J.: Biochem. Biophys. Res. Commun. *42*, 413 (1970)
127a. Liu, C.M., Johnson, M.J.: J. Bacteriol. *106*, 830 (1971)
128. Duppel, W., Lebeault, J.M., Coon, M.J.: Eur. J. Biochem. *36*, 583 (1973)
129. Gallo, M., Bertrand, J.C., Azoulay, E.: FEBS Lett. *19*, 45 (1971)
129a. Gallo, M. et al.: Biochim. Biophys. Acta *296*, 624 (1973)
130. Gallo, M., Roche, B., Azoulay, E.: Biochim. Biophys. Acta *419*, 425 (1976)
130a. Tittlebach, M., Rohde, H.G., Weide, H.: Z. Allg. Mikrobiol. *16*, 155 (1976)
131. Kawaguchi, K., Ishidate, K., Tagawa, K.: J. Biochem. *74*, 817 (1973)
131a. Wiseman, A., Lim, T.K., McCloud, C.: Biochem. Soc. Trans. *3*, 276 (1975)
131b. Wiseman, A., Woods, L.F.J.: Biochem. Soc. Trans. *5*, 1520 (1977)
132. Alexander, K.T.W., Mitropoulos, K.A., Gibbons, G.F.: Biochem. Biophys. Res. Commun. *60*, 460 (1974)
133. Wiseman, A., McCloud, C., Sin, T.K.: Biochem. Soc. Trans. *4*, 685 (1976)
133a. Lindenmayer, A., Smith, L.: Biochim. Biophys. Acta *93*, 445 (1964)
133b. Callen, D.F., Philpot, R.M.: Mutation Res. *45*, 309 (1977)
133c. Wiseman, A., Gondai, J.A.: Biochem. Soc. Trans. *3*, 33 (1975)
133d. Ishidate, K. et al.: J. Biochem. *65*, 375 (1969)
133e. Ishidate, K., Kawaguchi, K., Tagawa, K.: J. Biochem. *65*, 385 (1969)
134. Wiseman, A., Gondal, J.A., Sims, P.: Biochem. Soc. Trans. *3*, 278 (1975)
135. Wiseman, A., Jay, F., Gondal, J.A.: J. Sci. Food Agr. *26*, 539 (1975)
136. Yoshida, Y., Kumaoka, H.: J. Biochem. *78*, 785 (1975)
136a. Yoshida, Y., Kumaoka, H., Sato, R.: J. Biochem. *75*, 1201 (1974)
137. Yoshida, Y. et al.: Biochem. Biophys. Res. Commun. *78*, 1005 (1977)
138. Hedegaard, J., Gunsalus, I.C.: J. Biol. Chem. *240*, 4038 (1965)
139. Gunsalus, I.C., Pederson, T.C., Slinger, S.G.: Ann. Rev. Biochem. *44*, 377 (1975)
140. Katagiri, M., Ganguli, B.N., Gunsalus, I.C.: J. Biol. Chem. *243*, 3543 (1968)
141. Peterson, J.A., Basu, D., Coon, M.J.: J. Biol. Chem. *241*, 5162 (1966)
142. Peterson, J.A.: Arch. Biochem. Biophys. *144*, 678 (1971)
143. Yu, C.D., Gunsalus, I.C.: J. Biol. Chem. *249*, 94 (1974)
144. Yu, C.D., Gunsalus, I.C.: J. Biol. Chem. *249*, 102 (1974)
145. Dus, K. et al.: Biochem. Biophys. Res. Commun. *60*, 15 (1974)
146. Broadbent, D.A., Cartwright, N.S.: Microbios *3*, 113 (1971)

147. Cardini, G., Jurtshuk, P.: J. Biol. Chem. *243*, 6070 (1968)
148. Cardini, G., Jurtshuk, P.: J. Biol. Chem. *245*, 2789 (1970)
149. Appleby, C.A.: Biochim. Biophys. Acta *147*, 399 (1967)
150. Mazel, P., Hernandez, P.: Fed. Proc. *26*, 461 (1967)
151. Miura, Y., Fulco, A.J.: J. Biol. Chem. *249*, 1880 (1974)
152. Miura, Y., Fulco, A.J.: Biochim. Biophys. Acta *388*, 305 (1975)
153. Hare, R.S., Fulco, A.J.: Biochem. Biophys. Res. Commun. *65*, 665 (1975)
154. Orrenius, S., Ernster, L.: In: Molecular mechanisms of oxygen activation. Hayashi, O. (ed.), p. 215. New York: Academic Press 1974
155. Berg, A. et al.: Biochem. Biophys. Res. Commun. *66*, 1414 (1975)
156. Berg, A. et al.: J. Biol. Chem. *251*, 2831 (1976)
157. Breskvar, K., Hudnik-Plevnik, T.: J. Steroid Biochem. *9*, 131 (1978)
158. Wiseman, A., Lim, T.K.: Biochem. Soc. Trans. *3*, 974 (1975)
159. Edelson, J., McMullen, J.P.: Drug Metab. Disp. *5*, 185 (1977)
160. Smith, R.V.: In: Toxicology and nutrition – world review of toxicology and nutrition. Bourne, G.H. (ed.), Vol. 29, p. 60. Basel: S. Karger 1978
161. Scheline, R.R.: Pharmacol. Rev. *25*, 451 (1973)
162. Solomons, G.L.: Materials and methods in fermentation, p. 138. New York: Academic Press 1968
163. Weaver, E.A., Kenney, H.E., Wall, M.E.: Appl. Microbiol. *8*, 345 (1960)
164. Whitworth, D.A., Moo-Young, M., Viswanatha, T.: Biotechnol. Bioeng. *15*, 649 (1973)
165. Bryant, J.: In: Methods in microbiology. Norris, J.R., Ribbons, D.W. (eds.), Vol. 2, p. 187. New York: Academic Press 1970
166. Kulkarni, R.D., Goddard, E.D., Kanner, B.: J. Coll. Interfac. Sci. *59*, 468 (1977)
167. Elsworth, R.: Chem. Engineer. 63 (1972)
168. Calam, C.T.: In: Methods in microbiology. Norris, J.R., Ribbons, D.W. (eds.), Vol. 1, p. 255. New York: Academic Press 1969
169. Scarpa, A., Lindsay, J.D.: Eur. J. Biochem. *27*, 401 (1972)
170. Veldkamp, H.: In: Methods in microbiology. Norris, J.R., Ribbons, D.W. (eds.), Vol. 3A, p. 305. New York: Academic Press 1970
171. Bridges, J.W., Creaven, P.J., Williams, R.T.: Biochem. J. *96*, 872 (1965)
172. Raig, V.P., Ammon, R.: Arzneim.-Forsch. *22*, 1399 (1972)
173. Prough, R.A., Burke, M.D.: Arch. Biochem. Biophys. *170*, 160 (1975)
174. Davis, P.J., Jamieson, L.K., Smith, R.V.: Anal. Chem. *50*, 736 (1978)
175. Creaven, P.J., Parke, D.V.: Biochem. Pharmacol. *15*, 7 (1966)
176. Parke, D.V., Rahman, H.: Biochem. J. *119*, 53P (1970)
177. Bridges, J.W. et al.: Proc. Eur. Soc. Study Drug Toxicol. *15*, 98 (1973)
178. McPherson, F.J., Bridges, J.W., Parke, D.V.: Biochem. Soc. Trans. *2*, 618 (1974)
179. McPherson, F.J., Bridges, J.W., Parke, D.V.: Nature *252*, 488 (1974)
180. Parke, D.V.: In: Drug metabolism, from microbe to man. Parke, D.V., Smith, R.L. (eds.), p. 55. London: Taylor and Francis 1977
181. McPherson, F.J., Bridges, J.W., Parke, D.V.: Biochem. J. *154*, 773 (1976)
182. Billings, R.E., McMahon, R.E.: Mol. Pharmacol. *14*, 145 (1978)
183. Burke, M.D. et al.: Biochem. Soc. Trans. *5*, 1370 (1977)
184. Tong, S., Ioannides, C., Parke, D.V.: Biochem. Soc. Trans. *5*, 1372 (1977)
185. Atlas, S.A., Nebert, D.W.: Arch. Biochem. Biophys. *175*, 495 (1976)
186. Smith, R.V., Davis, P.J.: Unpublished
187. West, H.D. et al.: Arch. Biochem. Biophys. *60*, 14 (1956)
188. Raig, P., Ammon, R.: Arzneim.-Forsch. *22*, 1399 (1972)
189. Meyer, T., Scheline, R.R.: Acta Pharmacol. Toxicol. *39*, 419 (1976)
190. Meyer, T. et al.: Acta Pharmacol. Toxicol. *39*, 433 (1976)
191. Burke, M.D., Bridges, J.M.: Xenobiotica *5*, 357 (1975)
192. Gibson, D.T., University of Texas at Austin: Personal communication 1978

193. Ahmed, M., Focht, D.D.: Can. J. Microbiol. *19*, 47 (1972)
194. Ahmed, M., Focht, D.D.: Bull. Environ. Contam. Toxicol. *10*, 70 (1973)
195. Catelani, D., Sorlini, C., Treccani, V.: Experientia *27*, 1173 (1971)
196. Catelani, D., Colombi, A.: Biochem. J. *143*, 431 (1974)
197. Gibson, D.T. et al.: Biochem. Biophys. Res. Commun. *50*, 211 (1973)
198. Ohmori, T. et al.: Agr. Biol. Chem. *37*, 1599 (1973)
199. Lunt, D., Evans, W.C.: Biochem. J. *118*, 54P (1970)
200. Davis, P.J., Smith, R.V.: Abstr. Joint Central-Great Lakes Regional Meeting ACS, Indianapolis, IN, BIOL. 9, May 1978
201. Davis, P.J., Smith, R.V.: Abstr. First Joint Meeting Amer. Soc. Pharmacognosy, and the Phytochem. Soc. of North America, Stillwater, OK, August 1978

Major Chemical and Physical Features of Cellulosic Materials as Substrates for Enzymatic Hydrolysis

L. T. Fan, Yong-Hyun Lee, David H. Beardmore
Department of Chemical Engineering
Kansas State University
Manhattan, KS 66506, USA

This review emphasizes the structure and morphology of cellulose, which are pertinent to understanding the enzymatic hydrolysis of cellulose. Physical and chemical constraints on the susceptibility of cellulose to hydrolysis will be examined. In addition, the relationship between the capillary structure of cellulose fiber and enzymatic hydrolysis will be discussed.

1 Introduction

Glucose can be obtained from cellulose by acidic or enzymatic hydrolysis. The resulting glucose solution is useful as a substrate for production of single cell protein or alcohol. The production of glucose via cellulose degradation is by no means new. However, native crystalline cellulose is water insoluble, and its density and complexity make it very resistant to attack by the cellulolytic enzyme. To develop a commercial process

Table 1. Molecular weight and degree of polymerization of various celluloses[39]

Source	Molecular weight	Degree of polymerization
Native cellulose	600,000–1,500,000	3,500–10,000
Chemical cottons	80,000– 500,000	500– 3,000
Wood pulps	80,000– 340,000	500– 2,100
Rayon filament	57,000– 73,000	

of cellulose hydrolysis, the physical factors that make cellulose hydrolysis difficult should be identified and examined.

Comprehensive accounts of many aspects of cellulases and their application are available in the literature[45, 35, 19, 37, 50, 22, 52, 3, 18]. Papers on cellulases, presented at the 4th and 5th International Fermentation Symposia, have been published collectively[47, 13]. Reese and Mandels[38] reviewed enzymatic degradation of cellulose. Nisizawa[30] and Wood[53] published comprehensive reviews of the mode of action of cellulase. Reviews on various aspects of cellulase application also appeared in *Advances in Biochemical Engineering*[20, 21].

Comprehensive treatments of cellulose and its derivatives are available in a book by Bikales and Segal[5]. The journal *Cellulose Chemistry and Technology*[1] is a valuable source of information on physical, chemical, and technological aspects of cellulose and cellulosic materials. The most comprehensive sources of information on cellulose are the proceedings published after each of the international conferences on cellulose held every third year, often at Syracuse University, New York. The conference proceedings appear in the *Journal of Polymer Science, part C.*

2 Chemical and Molecular Structure of Cellulose

Cellulose is the most widely distributed skeletal polysaccharide, amounting to approximately 50% of the cell wall material of wood and plants. It is one of the three major components of wood and agricultural and municipal cellulosic wastes. The other two components are hemicellulose and lignin. Cellulose is a high molecular weight linear polymer composed of D-glucose residues, building blocks, joined by β-1,4-glucosidic bonds. Hydrolysis of these bonds yields mainly D-(+)-glucose as well as cellobiose that consists of two glucose residues. Table 1 shows typical cellulosic materials and their degrees of polymerization[39]. In native cellulose, up to 10,000 β-anhydroglucose residues are linked to form a long chain molecule. This means that the molecular weight of native cellulose is above 1.5 million. As the length of the anhydroglucose unit is 0.515 nm(=5.15 Å), the total length of the native cellulose molecule is about 5 μm. Cellulose of pulp and filter paper usually has a degree of polymerization ranging from 500 to 2100.

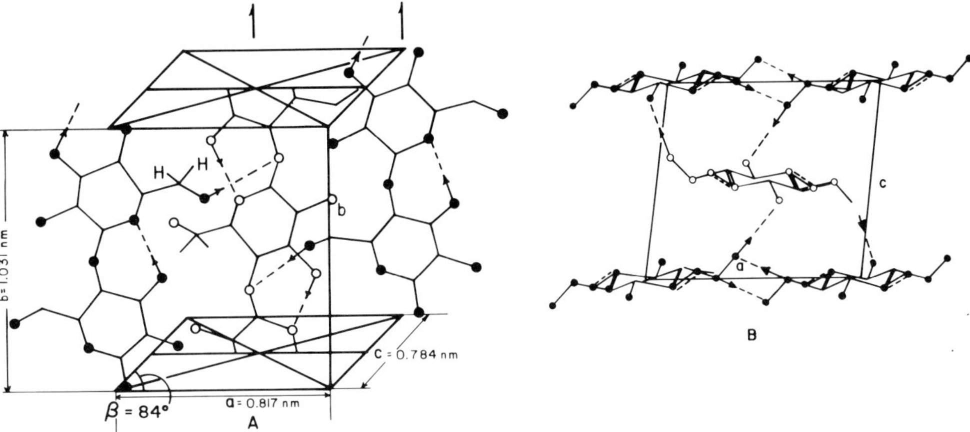

Fig. 1. Arrangement of cellulose molecules and hydrogen bonds in a cellulose 1 unit cell[27)]

In a cellulose chain molecule, the anhydroglucose units adopt the chair configuration with the hydroxyl groups in the equatorial and the hydrogen atoms in the axial positions. The conformational formula (chair form) of cellulose (poly-β-1,4-D-glucosan) is shown below.

Cellulose

Every other chain unit is rotated 180° around the main axis. The results in an unstrained linear configuration, with minimum steric hindrance[44)]. The glycosidic linkage acts as a functional group, and this, along with the hydroxyl groups, mainly determines the chemical properties of cellulose. All significant chemical reactions occur at these locations. It has been reported recently that the hydroxyl group in the 3-position is bound by an intramolecular hydrogen bond to the ring oxygen atom of the next chain unit[44)].

Cellulose molecules form a fibril, a threadlike long bundle of molecules, which is stabilized laterally by hydrogen bonding between hydroxyl groups of adjacent molecules. The molecular arrangement of this fibrillar bundle is sufficiently regular that cellulose exhibits a crystalline X-ray diffraction pattern. A schematic model of the native cellulose lattice was proposed by Meyer, Mark, and Misch in 1930. Figure 1 gives a spatial representation of the unit cell or crystal of cellulose developed by Liang and Marchessault[27)]. In this model, each unit cell contains four glucose residues. These include two in the center of the figure and one-fourth of each of the eight which are

Table 2. Stability of bonds in native cellulose[44)]

Dimensions	Length [nm]	Stability [kcal/mole]	Nature of bond
a	0.817	15	Hydrogen
b	1.031	50	Covalent
c	0.784	8	van der Waals

placed at the corners of a monoclinic cell. The corner residues are shared by each of the four unit cells which meet at the corners. The cellulose chain in the center of the unit cell runs in a direction opposite to that of the edges[44)].

The dimensions of the unit cell and the strength of different bonds are listed in Table 2. The length of the unit cell, which is 1.03 nm (1 nm = 10 Å), is the length of one of the repeating anhydrocellobiose units. Figure 1 B shows a projection of the unit cell on the ac-plane perpendicular to the direction of the b-axis. Notice that the shortest distance between atoms of neighboring chains of native cellulose is no more than 0.25 nm in the direction of the a-axis, which makes possible the formation of hydrogen bonds between adjacent chains. In the direction of the c-axis, the distance is much greater and molecular chains are attached to each other by van der Waal's forces only[44)].

Ward[51)] has pointed out that the consequence of the high degree of order in native cellulose is that neither a water molecule nor an enzyme molecule can enter the structure. Therefore, the native cellulose is inert in the digestive tract. However, the structure of acid- or alkali-swollen cellulose is open and is split relatively easily by cellulase.

3 Structure and Morphology of Cellulose Fibers

The chemical nature of cellulose, its physical and mechanical properties, and its fibrillary structure are deducible from its molecular structure. Like all hydrophilic linear polymers, individual cellulose molecules are linked together to form elementary fibrils or protofibrils, about 40 Å wide, 30 Å thick, and 100 Å long, in which the polymer chains are oriented in a parallel alignment and firmly bound together by numerous strong hydrogen bonds. An elementary fibril is the smallest structural unit of microfibrils and fibers, and a number of the elementary fibrils are aggregated into long slender bundles called microfibrils. In electron micrographs, the native cellulose microfibrils are usually seen as bundles in lamellae which contain an extremely large number of fibrillar units[33)].

A schematic representation of the cross section of a small lamella of microfibrils, proposed by Rånby[33)], is shown in Fig. 2. The crystalline region in which the linear molecules of the cellulose are bonded laterally by hydrogen bonds is characterized by the cellulose lattice which extends over the entire cross section of the microfibrils.

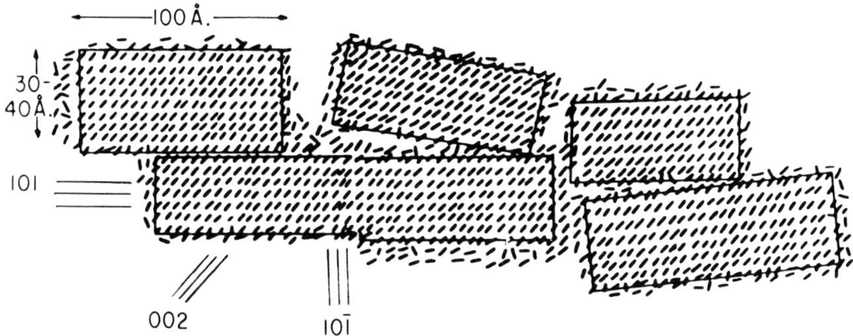

Fig. 2. Schematic representation of a lamella of cellulose microfibrils from the secondary wall of a plant[33)]

This crystalline region is bounded by a layer of cellulose molecules that exhibit various degrees of parallelism. The less ordered region is called the paracrystalline or amorphous region. The specific dimensions of microfibrils and spatial relationship between the crystalline and paracrystalline regions are still controversial subjects. The disordered region allows disintegration of the cellulose by hydrolysis into rod-like particles with aqueous, nonswelling, strong acid. Disordered areas of the microfibrils may be native or formed by mechanical forces, giving deformation beyond the limit of elastic recovery of the microfibrils. The length of a resultant particle after acid hydrolysis (micelles or microcrystals), which corresponds to the leveling-off degree of polymerization of the hydrocellulose, varies with the extent of the pretreatment of native cellulose[33)].

Cowling[11)] reviewed recent models of microfibril structure, shown in Fig. 3, from which we can make the following three observations: (a) The microfibril is about 50 x 100 Å in cross section and consists of a crystalline core of highly ordered cellulose molecules surrounded by a paracrystalline sheath. In cotton, this sheath contains mainly cellulose molecules, but, in wood, it also contains hemicellulose and lignin molecules. (b) The cellulose molecules are less well ordered at certain points along the length of a microfibril. (c) The cellulose molecules may exist in a folded chain lattice formed as a ribbon wound in a tight helix.

The fine structure of elementary fibrils (protofibrils) has been a subject of research for many years. Models of the various molecular arrangements of cellulose have been recently summarized by Chang[9)]. These models are shown in Fig. 4. The term "LODP" in Fig. 4 stands for leveling-off-degree-of-polymerization. There are two different types of models of the molecular arrangement of elementary fibrils. On the basis of the observation of chain folding in a linear synthetic polymer, models of cellulose fibrils involving folded chains (models c, d, and e) have been repeatedly suggested. On the other hand, models based on extended cellulose molecules without folding (models a and b) are also in the liberature. Chang[9)] has concluded that model c of Fig. 4 is supported by most of the known facts about cellulose.

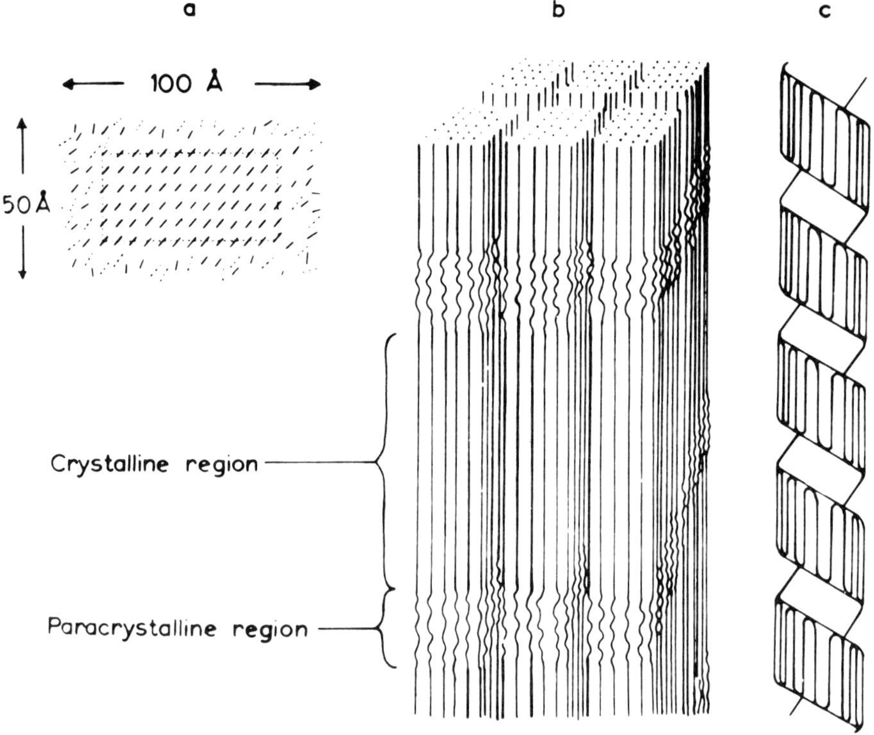

Fig. 3. Recent concepts of the structure of cellulose microfibrils[11)]

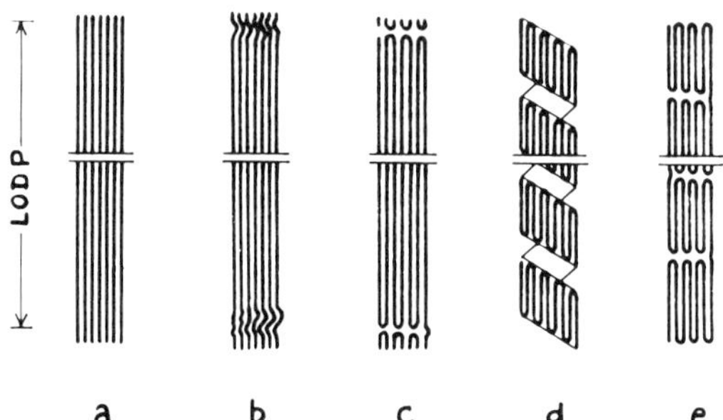

Fig. 4. Models of molecular arrangements in the protofibril: (a) Full extension and complete crystalline model (b) Fibrillized fringe micellar model (c) Fibrillized models of Ellefsen (d) Manley's planar zig-zag model (e) Chain folding with fold length much smaller than LODP[9)]

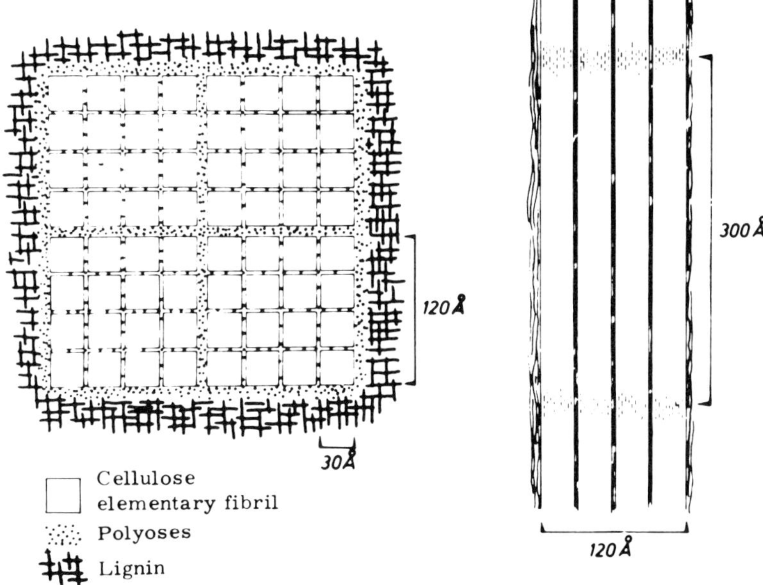

Fig. 5. Model of the ultrastructural organization of the cell wall components of wood[16)]

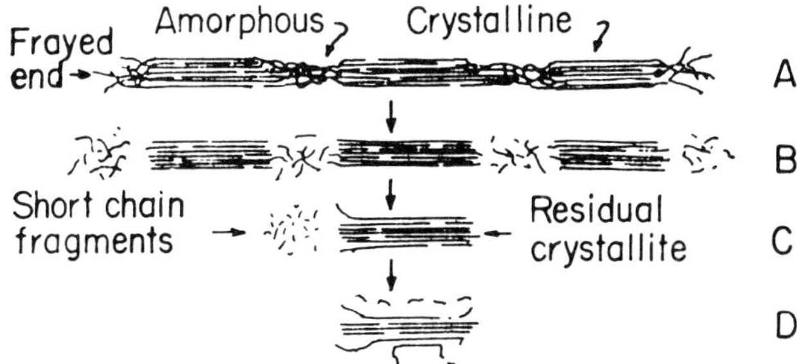

Fig. 6. Schematic summary of initial stages of dissimilation of a partially crystalline cellulose fibril: **(A)** Original cellulose fibril **(B)** Initial attack on amorphous region **(C)** Residual crystallite and digestion of short chain fragments **(D)** Attack on residual crystallite[30)]

Model c has been incorporated into a high-level structure, originally postulated by Fengel[15, 16)], which is shown in Fig. 5. This illustrates the ultrastructural organization of the cell wall components. The elementary fibrils are cemented together by polyoses such as hemicellulose to form a microfibril. This microfibril is surrounded by a lignin and polyose layer which protects the microfibril from enzyme action. Pretreatment of cellulosics, aimed at degrading these protective layers, is essential to rapid enzymatic hydrolysis.

The molecular structure of cellulose and the structures of an elementary fibril and a microfibril are important features from the standpoint of enzymatic degradation of cellulose. As mentioned previously, a cellulose fibril contains crystalline and amorphous regions. The proportion of crystalline or amorphous material has been estimated to range from 50 to 90%. Generally, about 70% crystallinity exists in native cellulose. Nisizawa[30] has depicted the different regions in a cellulose fibril at progressive stages of enzymatic degradation as shown in Fig. 6. In a partially crystalline cellulose fibril, the amorphous regions in the fringe micelles are first attacked, leading to an enrichment of crystalline regions. The crystalline regions are, in turn, gradually solubilized after loosening of their peripheral parts. Thus, the degradation of the substrate proceeds with mutual repetition of shortening of the chain length and subsequent loosening of the residual substrate. However, this scheme is highly idealized, considering the structural complexity of natural cellulose.

4 Physical and Chemical Constraints on the Susceptibility of Cellulose to Enzymatic Hydrolysis

The ability of cellulolytic microorganisms and that of cell free cellulolytic enzymes to degrade cellulose vary greatly with the nature of the substrate. Cowling and Brown[12] and Cowling[10, 11] have discussed comprehensively the influence of fiber structure on its susceptibility to enzymatic degradation. They have pointed out that the structural features of cellulosic materials which determine their susceptibility to enzymatic degradation include (1) the moisture content of the fiber; (2) the size and diffusibility of the cellulolytic enzymes and other reagent molecules involved in relation to the size and surface properties of the grown capillaries, and the space between microfibrils and the cellulose molecules in the amorphous region; (3) the degree of crystallinity of the cellulose; (4) the unit cell dimensions of cellulose; (5) the conformation and steric rigidity of the anhydroglucose units; (6) the degree of polymerization of the cellulose; (7) the nature of the substances with which the cellulose is associated; and (8) the nature, concentration, and distribution of substituent groups. Some of these structural features that are important in enzymatic hydrolysis of cellulose will be elaborated.

4.1 Degree of Water Swelling of Cellulose Fiber

The properties of wood and cellulose are profoundly affected by water. In partially dried cellulose, changes in the quantity of hygroscopically bound water induced by variations in the relative humidity of the surrounding atmosphere govern many mechanical and physical properties and the extent of swelling or shrinkage. Little or no swelling of wood or cellulose takes place in relatively nonpolar liquids, e.g., benzene, whereas more polar solvents, e.g., water, produce significant swelling[28].

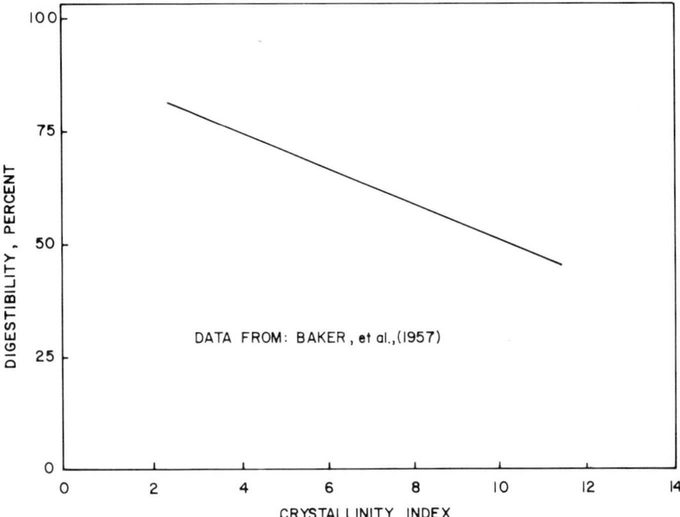

Fig. 7. Digestibility of cotton linters and wood cellulose against X-ray crystallinity index[14)]

The volumetric swelling of wood in water varies from 9 to 21.1% depending on the type of wood. The expanded capillary structure of water-swollen fibers may cause a significant increase in surface area of the cellulose fiber, and the total area of the swollen material may be as much as 100-fold greater than the area that results after drying from water[6)]. This opens up the fine structure so that the substrate is more accessible to cellulolytic enzyme and also facilitates diffusion of extracellular enzyme. Furthermore, molecules of water are added to the cellulose during hydrolytic cleavage of glycosidic links.

4.2 Crystallinity of Cellulose

The degree of crystallinity of cellulose is one of the most important structural parameters which affects the rate of enzymatic hydrolysis. The crystallinity of native cellulose was experimentally determined with an X-ray diffractometer using the focusing and transmission techniques by Segal et al.[43)]. They measured the intensity of the 002 interference and the amorphous scatter at $2\theta = 18°$. The fraction of crystalline material in the total cellulose was expressed in terms of an X-ray crystallinity index. Dunlap et al.[14)] has reanalyzed the data by Baker et al.[4)], as shown in Fig. 7, to examine the relationship between the cellulose crystallinity and digestibility. This figure indicates a linear inverse relationship between the crystallinity index and digestibility. Norkrans[32)] and Walseth[49)] showed that cellulolytic enzymes degrade the more readily accessible amorphous portions of regenerated cellulose, but are unable to attack the less accessible crystalline portions. They observed a significant increase in crystallinity during hydrolysis of cellulose with enzymes. As the highly accessible

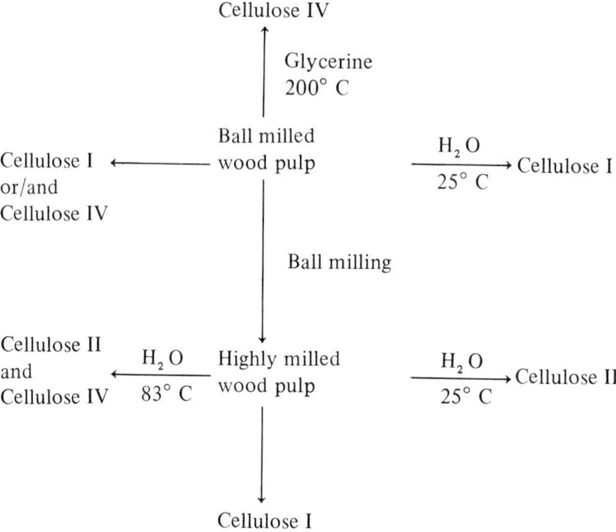

Fig. 8. Origins of the four crystal types of cellulose[23)]

materials become more crystalline, they show increasing resistance to further hydrolysis[40)].

Caulfield and Moore[8)] made somewhat different observations from those of Reese et al.[40)]. Their measurement of the degree of crystallinity of ball-milled cellulose before and after partial hydrolysis indicated that the mechanical action (ball milling) increases the susceptibility of both the amorphous and crystalline components of cellulose. They also observed that the enzyme digestibility of the crystalline component is enhanced to a greater extent by grinding than is the amorphous component. They concluded that the overall increase in digestibility is apparently a result of decreased particle size and increased available surface rather than a result of reduced crystallinity.

4.3 Molecular Structure of Cellulose

Cellulose exists in four recognized crystal structures designated as celluloses I, II, III, and IV[23)]. The sources of these four structural types are shown in Fig. 8. Rautela and King[34)] cultivated *Trichoderma viride* on the four different crystal forms and determined the activation energy for the enzymatic hydrolysis of each separate substrate. Their results showed that this organism produces an enzyme that yields the minimum activation energy for hydrolysis of a specific substrate. These results indicate that this fungus can synthesize an enzyme having active sites of a unique structure which accommodate a specific crystal lattice structure of an individual substrate.

King[24] has suggested that the greater resistance of crystalline than amorphous cellulose may not be due just to its physical inaccessibility to enzyme molecules, but also to the conformation and steric rigidity of the anhydrogluose units within the crystalline regions.

Rowland[42] studied the selective availability of hydroxyl groups in fibrous and crystalline cartons. It became evident from his study that the relative availability of hydroxyl groups on the most highly crystalline segments of the elementary fibril approached 1, 0, and 0.5 for $O_{(2)}H$, $O_{(3)}H$, and $O_{(6)}H$, respectively. These values of the theoretical relative availability on the surface of crystalline cellulose I result from the complete unavailability of $O_{(3)}H$ for reaction as a result of hydrogen bonding to $O_{(5')}$ and 50 % unavailability of $O_{(6)}H$ as a result of hydrogen bonding to $O_{(1'')}$.

4.4 Extraneous Materials – Lignin

The combination of lignin with partially crystalline cellulose that exists in wood constitutes one of nature's most chemically and biologically resistant materials. Many enzymes are prevented from degrading the cellulose in wood by the presence of lignin. Although the detailed chemical structure of lignin is reasonably well understood, the nature of the association between lignin and the wood polysaccharides still remains uncertain. A presently accepted view is that it is largely physical in nature, which means that lignin and amorphous cellulose form a mutually interpenetrating system of high polymers, but some covalent links exist between lignin and hemicellulose[17]. The relationship between the lignin content of untreated cellulosics and their respective digestibility has been presented by Millett et al.[29].

To degrade cellulose in wood, either the material must be partially delignified, the organisms must possess the ability to degrade cellulose and lignin, or the organisms must disrupt lignin's association with cellulose. Readers are referred to the recent reviews on the mechanisms for biological degradation of lignin by Kirk[25] and Rosenberg[41].

4.5 Presence of Substituent Groups

Substituted cellulose derivatives are formed by replacing the hydrogen of the primary and secondary hydroxyl groups of cellulose with reactive groups such as methyl, ethyl, hydroxyethyl, and carboxymethyl. The addition of these groups makes cellulose less crystalline and more soluble in water in proportion to the degree of substitution (DS) and the solvating capacity of the substituent groups. The DS at which complete solubility is attained ranges from 0.5 to 0.7 depending on the solvating capacity of the substituents and the degree of polymerization of the cellulose[12].

Susceptibility of substituted cellulose derivatives to enzymatic hydrolysis increases as they become more water soluble and less crystalline up to the complete solubility point. After this point, the susceptibility decreases with increasing DS until complete immunity to enzyme action results. This usually occurs at a DS somewhat greater than 1.0. Substituent groups of large molecular dimensions are more effective in contributing resistance to enzymatic degradation than small groups[36].

5 Capillary Structure of Cellulose Fibers and Enzymatic Hydrolysis

The susceptibility of cellulose to enzymatic hydrolysis is determined largely by its accessibility to a cellulolytic enzyme. Direct physical contact between the enzyme and the substrate molecules of cellulose is a prerequisite to hydrolysis. Since the cellulose is an insoluble and structurally complex substrate, this contact can be achieved only by diffusion of the enzymes into the complex structural matrix of the cellulose. Any structural feature that limits the accessibility of the cellulose to enzymes will diminish the susceptibility of cellulose to hydrolysis. The influence of change in the structure and composition of a variety of cellulosic substrates upon their susceptibility to enzymatic attack has been described in the preceding section.

While a good correlation has often been obtained between a particular structural feature and the hydrolysis rate of a substrate, a great deal of ambiguity remains. The lack of agreement between different sets of data can be illustrated by the following examples. Lee[26] has found that the most reactive regenerated cellulose is also the most crystalline, in contrast to most reports. This implies that the degree of swelling, not the crystallinity of the substrate, is the most important factor, because a higher degree of swelling increases the pore size above a certain critical value which is necessary to allow enzyme molecule to penetrate the substrate. Another example of the confusion arises in equating the accessibility of the cellulose to its amorphous fraction[46]. A highly amorphous substrate may be slow to hydrolyze because the amorphous material in the wood is accessible to water and hydrogen ions but is completely inaccessible to an enzyme. These two examples emphasize the importance of capillary structure and enzyme reactivity to hydrolysis.

In this section, the most significant structural feature will be considered. This is the capillary structure of cellulose relative to the size and diffusivity of each cellulolytic enzyme. Direct physical contact between the enzyme and its substrate produces an enzyme-substrate complex which then breaks down to yield the product of the reaction. It is expected, therefore, that the rate of reaction should be a function of the surface area of cellulose fibers as defined by the size, shape, and surface properties of the microscopic capillaries within the fiber in relation to the size, shape, and diffusibility of the cellulolytic enzyme itself.

5.1 Capillary Structure of Cellulose Fibers

The capillary voids in cellulosic fibers include (1) gross capillaries such as the cell lumina, pit apertures, and pit-membrane pores that are visible in the light microscope and are in the range between 200 Å and 10 or more microns in diameter and (2) cell

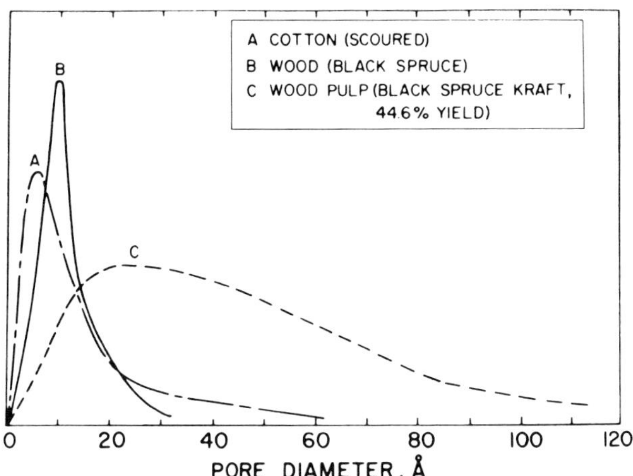

Fig. 9. Frequency distribution for cell wall capillaries[12]

wall capillaries such as the spaces among microfibrils and those among cellulose molecules in the amorphous regions, which are less than 200 Å in diameter[12]. The total surface area exposed is the gross capillaries in quite large, approximately 2 x 10^3 cm^2 per gram of wood or cotton. It is, however, several orders of magnitude smaller than the total surface area exposed within the cell wall capillaries, which is approximately 3 x 10^6 cm^2 g^{-1} [10]. The area exposed on the gross capillary surfaces of one gram of wood or cotton is sufficient to accommodate about 3 x 10^{15} randomly oriented enzyme molecules 200 x 35 Å in size, which is equivalent approximately to 3 mg of enzyme protein per gram of wood or cotton. It should be noted that the hemicellulose and lignin occupy the spaces among the microfibrils and cellulose molecules in the amorphous regions. Cutting or grinding the wood would produce a considerable amount of additional contacting surface between the cellulose and enzyme. If vibratory ball milling or other treatment do not increase reactivity, it must be due to the changes within the cell wall itself rather than to the size of the cell wall fragment.

The dimensions of the cellulose capillaries have been determined recently using the solute exclusion technique by Aggebrandt and Samuelson[2], Stone and his coworkers[46], Nelson and Oliver[31], and Van Dyke[48]. Figure 9 shows a frequency distribution of capillary dimensions in water-swollen wood, cotton, and wood pulp. The median and maximum dimensions of cell wall capillaries in the three materials are about 5 and 40 Å for water-swollen wood, about 10 and 75 Å for water-swollen cotton, and about 45 and 110 Å for water-swollen wood pulp[12]. The substantial increase in the median and maximum pore sizes during pulping results, in part, from removal of lignin and hemicellulose from cell wall capillaries.

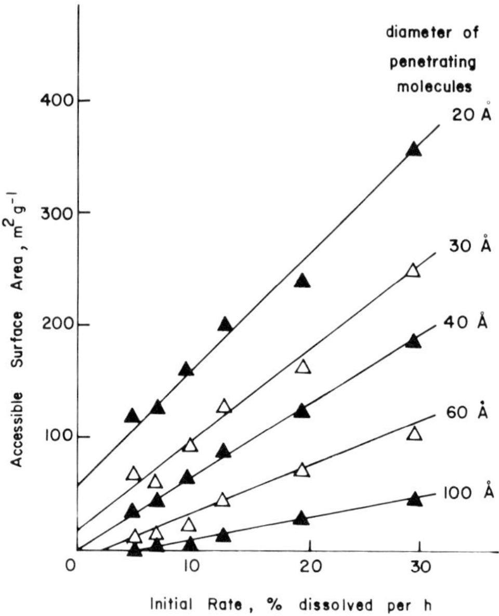

Fig. 10. Relationship between the initial rate of reaction and the surface area of cellulose accessible to molecules of different sizes[46)]

5.2 Penetration of Enzyme into Capillaries in Natural Fibers

Cellulolytic enzyme molecules range from 24 to 77 Å in diameter, with an average of 59 Å[11)]. Thus, these molecules would be expected to diffuse readily within the gross capillaries and act on cellulose molecules exposed on the surface of these capillaries. However, only a small fraction of the cell wall capillaries in water-swollen wood and cotton fibers is sufficiently large to permit penetration of most of the cellulolytic enzyme molecules. For this reason, it is likely that the cellulolytic enzymes are physically excluded from all but the largest cell wall capillaries. For them to gain access to these regions, the capillaries must be enlarged.

5.3 Digestibility as a Function of Pore Size Distribution and Surface Area of Cellulose Fibers

Stone et al.[46)] measured the pore volume using the polymer exclusion technique and calculated the surface area within the swollen fibers accessible to all the molecules. The substrate reacted with a commercial cellulase enzyme preparation, and the initial rate of reaction was compared with the accessibility of the substrate to molecules of various sizes. A linear relationship was found between the initial reaction rate and the surface area within the cellulose gel which was accessible to a molecule of 40 Å diameter. Figure 10 shows the relationship between the initial rate of reaction and the surface area of cellulose accessible to molecules of different sizes. It appears that the digestibility of cellulose with different degrees of swelling is directly proportional to the accessibility of molecules of 20–40 Å in diameter, which might correspond to the size of cellulase molecules.

6 Conclusions

Cellulose is a linear homopolymer of anhydroglucose units linked together with β-1,4-glucosidic bonds. Cellulose molecules are linked together to form elementary fibrils, or protofibrils of about 40 Å wide and 30 Å thick. The fine structure of elementary fibrils has been a subject of research for many years. Two major schools of thought exist. Based on the observation of chain folding in a linear synthetic polymer, models of cellulose fibrils involving folded chains have been repeatedly proposed. On the other hand, models based on extended cellulose molecules without folding are also in the literature. However, the subject is still controversial. The aggregates of elementary fibrils, of essentially an infinite length and a width of approximately 250 Å, are called microfibrils, and they contain zones of less crystalline order between the elementary fibrils. The heavily lignified cell wall surrounding the fiber reveals the cementing role of lignin and indicates its possible hindrance to cellulose hydrolysis. To remove this lignin layer, intensive mechanical grinding and chemical treatment are often employed in pretreatment of cellulosic materials prior to hydrolysis by either acids or enzymes.

Cellulose contains crystalline and amorphous regions. The proportion of crystalline materials is estimated to be 50–90%, and the remainder as amorphous material. The different regions in cellulose exhibit different susceptibilities at progressive stages of enzymatic degradation. The structural features of cellulosic materials which determine their susceptibility to enzymatic degradation include (1) the degree of swelling by water, (2) the crystallinity of cellulose, (3) the molecular arrangement of cellulose, (4) the presence of extraneous material such as lignin, and (5) the presence of substituent groups. However, the most important structural feature which affects enzymatic hydrolysis may be the capillary structure of the cellulose fiber. This is because the susceptibility of cellulose to hydrolysis is determined largely by its accessibility to cellulolytic enzyme; direct physical contact between the reagents and the substrate molecules of cellulose is prerequisite to hydrolysis. Since the cellulose is an insoluble and structurally complex substrate, this contact can be achieved only by diffusion of the enzyme into the complex structural matrix of cellulose. Any structural feature that limits the accessibility of cellulose to enzymes will diminish its susceptibility to hydrolysis. Therefore, the rate of reaction should be a function of the surface area of the cellulose which is accessible to the enzyme. It is reported that a linear relationship exists between the initial reaction rate and the surface area within cellulose which is accessible to a molecule of the enzyme size.

The least known aspect of the enzymatic hydrolysis of cellulose may be the effect of the major structural features of cellulose on the hydrolysis rate. This relationship should be further examined until it is fully understood. This knowledge will aid in the development of an economical and effective pretreatment which can modify the cellulose structure to give the optimal hydrolysis rate.

Acknowledgments

This is Department of Chemical Engineering contribution No. 79-118-R, Agricultural Experiment Station, Kansas State University.

7 References

1. Academy of the Socialist Republic of Rumania (ed.): Cellulose Chem. Tech. Rumania: Bucharest
2. Aggebrandt, L.G., Samuelson, O.J.: J. Appl. Polym. Sci. *8*, 2801 (1964)
3. Bailey, M., Enari, T.M., Linko, M. (eds.): Symposium on Enzymatic Hydrolysis of Cellulose. SITRA, Helsinki 1975
4. Baker, T.L. et al.: J. Anim. Sci. *18*, 655 (1959)
5. Bikales, N.M., Segal, L. (eds.).: Cellulose and cellulose derivatives, Vol. 5, Part V, 2nd ed. New York: Wiley Interscience 1971
6. Browning, B.L.: In: The chemistry of wood. Browning, B.L. (ed.), p. 429. New York: Interscience 1963
7. Callihan, C.D.: ACS Symposium Series *10*, p. 27. ACS, Wash., D.C. 1975
8. Caulfield, D.F., Moore, W.E.: Wood Sci. *6*, 375 (1974)
9. Chang, W.: J. Polym. Sci.: Part C, No. 36, 343 (1971)
10. Cowling, E.B.: In: Advances in enzymic hydrolysis of cellulose and related material. Reese, E.T. (ed.), p. 1. New York: Pergamon Press 1963
11. Cowling, E.B.: Biotechnol. Bioeng. Symp. No. 5, 163 (1975)
12. Cowling, E.B., Brown, W.: Adv. Chem. Ser. *95*, 152 (1969)
13. Dellweg, H. (ed.): Proc. Vth Int. Ferm. Symp. Berlin 1976
14. Dunlap, C.E., Thomson, J., Chiang, L.C.: AIChE Symp. Ser. No. 158, *72*, 58 (1976)
15. Fengel, D.: TAPPI, *53*, 497 (1970)
16. Fengel, D.: J. Polym. Sci.: Part C, No. 36, 383 (1971)
17. Wardrop, A.B.: In: Lignins. Sarkanen, K.V., Ludwig, C.H. (eds.), p. 19. New York: Wiley-Interscience 1971
18. Gaden, Jr., E.L. et al. (eds.): Enzymatic conversion of cellulosic materials: technology and applications. New York: Interscience 1976
19. Gascoigne, J.A., Gascoigne, M.M.: Biological degradation of cellulose. London: Butterworths 1960
20. Ghose, T.K., Fiechter, A., Blakebrough, N. (eds.): Adv. in Biochem. Eng., Vol. 5, Berlin: Springer 1977
21. Ghose, T.K., Fiechter, A., Blakebrough, N. (eds.): Adv. in Biochem. Eng., Vol. 6, Berlin: Springer 1977
22. Hajny, G.J., Reese, E.T. (eds.): Celluloses and their applications. Adv. in Chem. Ser. *95*, ACS, Wash., D.C., 1969
23. Howsmon, J.A., Marchessault, R.H.: J. Appl. Polym. Sci. *1*, 313 (1959)
24. King, K.W.: Va. Agric. Expt. Sta. Tech. Bull., 154 (1961)

25. Kirk, T.K.: Biotechnol. Bioeng. Symp. No. 5, 139 (1975)
26. Lee, Jr., G.F.: Physical factors affecting enzymatic hydrolysis of cellulose. M.S. Thesis, Syracuse Univ., Syracuse, N.Y. 1966
27. Liang, C.Y., Marchessault, R.H.: J. Polym. Sci. *37*, 385 (1959)
28. Merchant, M.V.: TAPPI *40*, 771 (1957)
29. Millett, M.A., Baker, A.J., Satter, L.D.: Biotechnol. Bioeng. Symp. *5*, 193 (1975)
30. Nisizawa, K.: J. Ferment. Technol. *51*, 267 (1973)
31. Nelson, R., Oliver, D.W.: J. Polym. Sci.: Part C, No. 36, 305 (1971)
32. Norkrans, B.: Physical Plant *3*, 75 (1950)
33. Rånby, B.: Adv. Chem. Ser. *95*, 134 (1969)
34. Rautela, G.S., King, K.W.: Arch. Biochem. Biophys. *123*, 589 (1968)
35. Ray, D.L. (ed.): Marine boring and fouling organisms. Univ. of Wash. Press, Seattle 1959
36. Reese, E.T.: Ind. Eng. Chem. *49*, 89 (1957)
37. Reese, E.T. (ed.): Advances in enzymatic hydrolysis of cellulose and related materials. New York: Pergamon Press 1963
38. Reese, E.T., Mandels, M.: In: Cellulose and cellulose derivatives. Bikales, N.M., Segal, L. (eds.), p. 1079. New York: John Wiley 1971
39. Reese, E.T., Mandels, M., Weiss, A.H.: In: Adv. in Biochem. Eng. Ghose, T.K., Fiechter, A., Blakebrough, N. (eds.), Vol. 2, p. 181. Berlin: Springer 1972
40. Reese, E.T., Segal, L., Tripp, V.M.: Text. Res. J. *27*, 626 (1957)
41. Rosenberg, S.L.: Proc. AIChE 81st National Meeting, Kansas City, Mo. 1976
42. Rowland, S.P.: Biotechnol. Bioeng. Symp. No. 5, 183 (1975)
43. Segal, L. et al.: Text. Res. J. *29*, 786 (1959)
44. Sihtola, H., Neimo, L.: In: Symp. on enz. hyd. of cellulose. Bailey, M., Enari, T.M., Linko, M. (eds.), p. 9. SITRA, Helsinki 1975
45. Siu, R.G.H. (ed.): Microbial decomposition of cellulose, Reinhold, New York: 1951
46. Stone, J.E. et al.: Adv. Chem. Ser. *95*, 219 (1969)
47. Terui, G. (ed.): Fermentation technology today: Proc. IVth Internat. Ferm. Symp., Soc. Ferm. Technol., Osaka 1972
48. Van Dyke, Jr., B.H.: Enzymatic hydrolysis of cellulose – a kinetic study. Ph.D. Dissertation, MIT, Cambridge, Mass. 1972
49. Walseth, C.S.: TAPPI *35*, 233 (1952)
50. Walters, A.H., Elphick, J.J. (eds.): Biodeterioration of materials. New York: Elsevie 1968
51. Ward, Jr., K.: Symp. on foods: carbohydrates and their roles. Schultz, H.W., Cain, R.F., Wralstad, E.W. (eds.), p. 55, AVI Pub. Co., Westport, Conn. 1969
52. Wilke, C.R. (ed.): Cellulose as a chemical and energy resource. Biotechnol. Bioeng. Symp. No. 5, New York: Interscience 1975
53. Wood, T.M.: Biotechnol. Bioeng. Symp. No. 5, 111 (1975)

Recent Developments in the Large Scale Cultivation of Animal Cells in Monolayers

Raymond E. Spier
Animal Virus Research Institute
Pirbright, Woking, Surrey, England

The history of monolayer cell culture is reviewed and the potential product applications for cells grown in this way outlined. Recent developments in understanding the basic components of this system and the way in which cells interact with the surface on which they grow are considered. The general features of the different types of equipment which have the potential of producing such cells on the industrial scale are reviewed. A discussion of the particular equipment types follows and leads to the conclusions that no one system has yet been perfected and that most potential for development resides in the glass sphere packed beds and/or microcarrier systems.

1 Introduction

The recent interest in new methods of providing surface for the growth of animal cells calls for evaluation. Conventially, those animal cells which require a surface to grow upon are cultivated on the walls of glass bottles. The bottles generally used for this purpose may be held in stationary racks or they may be set upon rollers and slowly rotated. Using such bottles, often by the thousand (Fig. 1), many successful processes for the production of viral vaccines have been developed. However, such processes, relying upon a multiplicity of bottles (multiple processes), are inefficient and have little potential for improvement so that efforts are presently directed towards consolidating the surface provided by many bottles into one container. The resulting unit processes for the production of animal cells and the materials which can be generated from such cells form the subject matter of this chapter.

It is evident that there are many advantages to be gained from the transition from a multiple process to a unit process. Each development which purports to achieve this will be evaluated by the extent to which the expected advantages are realised in practice. Apart from the technical advantages of unit process systems, further reasons for developing very large scale (1000 l) systems come from (a) the requirement to produce interferon from human diploid fibroblasts in sufficient quantities for general therapeutic use; (b) the realisation that potent vaccines against certain strains of foot-and-mouth disease (FMD) virus require the use of surface-dependent Baby Hamster Kidney cells (BHK 21 C13) and (c) a growing need to produce larger quantities of the inactivated polio vaccine.

Developments in new technology are also dependent on advances in cell biology. The quality of the media, sera and cells are important variables in cell production systems and often determine the applicability of a particular unit process system. The interaction between these biological parameters and the equipment in which they express themselves is also examined below.

2 Background

Animal cell culture developed into an active area of investigation at the end of the 19th century. Reviews by Wilmer[1), Paul[2), and Weymouth[3) describe the development of culture media and the early investigations on the methods of growing animal cells outside the body. During the decade 1950–1960 considerable progress was made in the formulation of media. Earle and his coworkers in 1951[4) and later in 1953/4[5) began to consider methods for the production of "massive cultures" on "three-dimensional substrates". In this work, folded cellophane sheets and packed beds made from either a mass of 1/8 inch diameter Pyrex glass helices or 6 mm diameter glass Raschig rings[6) were used. At this early date it was clear that certain physical parameters would significantly affect cell growth. The velocity of rotation of a tube, the linear rate of flow of medium over the inner surface of the tube and the

Fig. 1. One of four incubators each holding 7,200 individual roller bottles (with thanks to Dr. J.F. Panina, Istituto Zooprofilatico Sperimentale de l'Emilia at Lambardia, Brescia, Italy)

initial cell inoculum concentration were considered important. They concluded that the cellophane sheets lacked sufficient structural rigidity but that a matrix made from "shaped glass particulate forms" was best. The scale-up of such cultures from 100 l appeared to be "entirely practical".

After a gap of 8 years, McCoy et al.[')] followed up Earle's work and showed that cells would replicate in a static bed of glass helices when medium was perfused through them. This system consisted of 12 individual 40 ml beds of helices fed from a common medium reservoir. A cell yield of 1 x 10^6 cells per ml could be obtained.

The year 1967 marks the turning point in the development of systems for the large-scale propagation of animal cells. In this year Molin and Hedén[8] presented their paper on the cultivation of human diploid cells on titanium discs and van Wezel[9] described the growth of primary cells and cell strains on DEAE Sephadex A50 micro-carriers. Following these developments, one or more new systems for the bulk culture of animal cells have been described each year. The remainder of this chapter will consider these systems.

3 Products Derived from Surface-dependent Animal Cells Grown in Tissue Culture

A great variety of products can be made from animal cells (Table 1). They may be divided into five major groups: a) viral vaccines, b) interferon, c) hormones, d) immunological reagents and e) cellular biochemicals such as enzymes or materials which have been specifically transformed by cultivated cells. At the present time the viral vaccine application is the most widely used.

The factors which control the use of products derived from animal cells are complex. In most countries regulations have been formulated to protect the population against the use of ineffective and unsafe agents. Such regulations are generally administered by an independent agency funded by the government. Thus every new system for generating a product from a tissue culture cell is examined in great detail by the regulatory agency of the country in which the product is to be used. For this reason it is important to understand how such regulatory agencies work and hence to appreciate the motives behind the drive for the production of interferon from surface-dependent fibroblastic cells.

Regulatory agencies. Each product which is made from a cultivated animal cell must be granted a licence prior to sale in the society at large. Such licences are granted on the basis of the demonstrated efficacy and safety of the product. There are no defined rules by which either the efficacy or safety of a product may be established. Each product from each manufacturer is examined by members of the licensing authority in consultation with the production staff of the manufacturing company. In such meetings tests to define the quality of the product are agreed upon. The results of the tests are also discussed and, if necessary, further tests are proposed until the licensing authority is satisfied that the product is safe and efficacious.

In addition to the specific examination of a particular product there is a series of guidelines governing many aspects of the facilities in which a product for human use may be made[12, 14] and also the type of cell from which such products may be derived[11]. Two cell types may be used for generating products for human use. The first is the primary cell produced by the trypsinisation of a specified excised tissue derived from an animal which has been shown to be in good health and the second is from a highly defined and controlled human diploid cell line. Cells of the latter type must (a) exhibit a karyology which is almost identical with the parent tissue, (b) exhibit a decreased or zero growth capability after about 40–50 passages in

Table 1. Products made from animal cells grown in monolayers

Product		Licenced	Ref.
1. Vaccines			
– polio	L[a]	yes	12, 36
– polio	D	yes	12, 24
– FMD	D	yes	35
– mumps	L	yes	13, 30
– measles	L	yes	13, 29
– rubella	L	yes	13, 28
– Mareks	L	yes	34
– yellow fever	L	yes	12
– rabies	D	yes	32, 33
– others			12, 25
2. Interferon			26, 27, 41
3. Hormones			31, 37, 40,41
4. Immunological reagents			38, 41
5. Cellular biochemicals			
– chalones			39
– insulin			41
– enzymes			41
– plasminogen			42
– plasminogen activator			43

[a]L: uninactivated vaccine; D: inactivated vaccine

tissue culture (c) be free from adventitious contamination by mycoplasma, bacteria, fungi, viroids or other transforming principles (e) be fully controlled during each production run by demonstrating normal behaviour. Such regulations are not applied to products for use in veterinary applications where BHK cells grown in suspension may be used. However, recommendations for greater control over the way in which such suspension cells are used are presently under consideration[10)].

Apart from the detailed examination of each new product, the regulatory agency may take into consideration the relationship between the new product and a product which has been available for many years. If the established product has performed well and safely, then, providing the changes incurred in the production of the new product are small and not contentious, a licence may be granted quite rapidly. Such a situation occurred when the process to produce mumps, measles and rubella virus vaccines was adapted from a roller bottle system to a unit process system in which the cells were grown on titanium discs[13)].

The agency is also influenced by the balance of risks against gains with regard to the use of a particular product. For example, before the advent of polio vaccines the number of polio cases was about 100–200 per million population[15)]. After vaccination at first with inactivated vaccines and then with live attenuated virus vaccines the number of cases dropped to 1 case per 4.5 million doses of live virus vaccine administered[16, 23)]. Now that the overall level of polio is low the risks associated with the live virus vaccine have become significant and there is a growing movement to replace it with the inactivated vaccine which has not been associated with any case of polio[17, 23].

The situation with regard to the use of interferon is still under consideration. At present it is possible to produce useful quantities of interferon from a lymphoblastoid cell line which has been transformed by an oncogenic agent, Epstein-Barr virus[18]. However, in spite of the various purification steps (immunoadsorbent columns, chloroform precipitation and pelleting from 94% ethanol at pH 4), a licence for the general use of this product has not yet been granted. For this reason the alternative methods of producing interferon from cell lines which have been approved for the production of material fit for human use have received considerable attention. Such approved cell lines (WI38[19] MRC5[20] IMR90[21]), require a surface for growth and as the product is required in very large quantities the development of a unit process technology becomes virtually essential. (A typical treatment could consist of 3–4 doses a day where each dose contains 1×10^6 units of interferon which is produced by 30–60 ml of tissue culture fluid.)

Foot-and-mouth disease virus. While the impetus to develop large-scale equipment for the cultivation of monolayer cells for interferon has come from the difficulty of obtaining a licence to use lymphoblastoid interferon, the reason for developing similar equipment for the production of FMD vaccine has come from the requirement to produce strains of FMD virus which can only be produced to immunogenic levels in BHK monolayer cells[22]. Large quantities of such vaccines are required in the big cattle-producing countries where animals are vaccinated two or more times per year.

Polio vaccine. The current movement towards the use of inactivated polio vaccines is based on four reasons[23]: (a) the balance of the risks against the gains has changed considerably since vaccination against polio was started (see above), (b) the attenuated virus vaccine is less effective in those areas of the world where poorly nourished children may be either infected with strains of enterovirus before vaccination or have inhibitors of the immunogenic reaction present in their intestines (c) the cost difference between the inactivated and attenuated live virus vaccines has decreased (d) it has been shown that injection of inactivated vaccine is not significantly more difficult to administer than the oral ingestion of a sugar cube. The production problems which this change causes can be seen in the need to *concentrate* the tissue culture fluid to prepare the inactivated polio vaccine[24] while the live vaccine is made from *diluted* tissue culture fluid.

Other virus vaccines. The inactivated vaccines referred to above are made in very large quantities. Other vaccines made from attenuated live virus preparations are often diluted before vaccine formulation. It is thus possible to produce 20×10^6 doses of vaccine each year from about 2000 bottle cultures where each culture yields 100 ml of virus at 100 times the use concentration. The reasons to convert such a multiple process into a unit process are not founded on the impracticability of the multiple operation but rather from the increase in the efficiency of the unit operation which reduces the overall costs of production.

4 The Basic Components of the Tissue Culture Systems

The successful operation of equipment of advanced design is critically dependent on the quality of the materials of which it is made and the quality of the materials which are fed into it. Current research has increased the number of possible alternatives in both of these areas. In this section recent developments in medium formulation, cell biology and the relationship between a cell and the surface to which it attaches itself and on which it grows and divides will be examined.

4.1 Medium Composition

Media of different degrees of complexity have been formulated. They consist of mixtures of inorganic salts, amino-acids, vitamins, nucleotides and low molecular weight growth factors such as hormones, steroids and fatty acids. Most media for the growth of primary cells and diploid cells require serum and are often further fortified with complex mixtures of peptones, tryptic digests and albumin hydrolysates[3]. Current investigations have sought to discover the effect of amino-acid composition on the growth and longevity of diploid fibroblasts and ways of controlling or eliminating the serum component of the medium.

Amino-acid composition. During investigations on the amino-acids necessary to support the growth of human diploid fibroblasts (HDF), it was found that, of the 13 amino acids required for the formulation of Eagle's minimum essential medium (MEM), only two amino-acids glutamine and cystine[45] or glutamine and cysteine[44] in addition to the salts, 10% calf serum and folic acid[44] would suffice. The choice of antibiotic mixture could also influence cell growth and longevity[50].

Serum quality. Perhaps the most important determinant of the growth-promoting capability of the medium is the quality and quantity of the serum used. Many tests have been prepared to select batches of sera for use in critical tissue-culture operations[46–48, 159]. In addition, much work has been devoted to analysing the serum in order to isolate the growth-promoting factors[57] (see also reviews[49, 50, 58]) and whilst many fractions have been found to be active it has not been possible to fully characterise the growth promoters.

Many investigators have taken the alternative approach of formulating a medium which does not require serum. The potential advantages of such a medium would be (a) low cost (b) heat sterilisability (c) chemical definability (d) ability to produce a virus in the absence of antibodies to that virus (e) consistency of quality (f) reduction of allergic reactions (g) reduction of exogenous contamination by mycoplasma and virus strains[51–53]. However, serum-free media do not support the growth of primary or HDF cells and may only be used for cells which grow in suspension[54, 55] or are transformed (see below).

Additional growth factors have been sought in (a) medium in which cells have been grown (conditioned medium)[50, 59], (b) factors discharged by X-radiated cells[59] or (c) different cell types[50].

4.2 Cell Quality

The factors which control cell quality are dependent on the type of cell under consideration. The three main cell types used for vaccine or interferon production are (a) primary (or secondary) cells, (b) diploid cell lines and (c) established cell lines. These cell types and others are defined in detail in Federoff[60].

Primary cells. These cells are generally obtained from the trypsinisation of a decapitated avian embryo or a tissue excised from a healthy animal. In the former case the age of the embryo and the treatment of the egg from the moment it is laid are important variables. The age of the embryo determines the cell yield though older embryos may be difficult to handle because of feathers. In order to fully control the quality of the embryo it is necessary to specify in detail the source of the eggs (specific pathogen-free hens) and the processing of the eggs from the moment of laying. It is not uncommon for eggs obtained from commericial sources to be held at ambient temperatures for several days before beginning the incubation at 37 °C.

Primary cells derived from green monkey kidneys are commonly used for polio vaccine preparation. The green monkeys must be free from the Simian papovavirus 40 (SV40)[61] and are very expensive (£ 150 per monkey). For this reason the technique of van Wezel[62] which doubles the yield of cells which can be obtained, by perfusing the kidney with trypsin while it is still in the body, is an important development.

Other primary cells such as rabbit kidney (for Rubella, Vaccinia) calf kidney (for measles in Japan) are also used for human vaccines[64]. The range of primary cells which can be used for FMD virus production is very wide and includes bovine and porcine kidneys, goat heart and foetal camel lung, skin and kidney cells[63].

Diploid cells. In view of the difficulties of obtaining adequate quantities of primary green monkey kidney cells at a reasonable price and the advantages of using a prescreened cell culture, polia vaccine production in WI38 and MRC5 diploid cell lines has also been licenced. Other vaccines may also be prepared in these cell lines[65]. In addition to vaccines, interferon may be prepared from such cells.

There is little doubt that diploid cells vary in their growth properties and ability to generate useful products. Even well-characterised diploid cell lines behave very differently in the hands of different investigators. Such differences may take the form of differences in growth rate, longevity and ability to produce interferon[27]. In the latter case a 6-fold difference could be seen between different cell lines and the yield of interferon from cells between 16–20 population doublings was, on average, more than 12 times that from the same cells between 55–60 population doublings. In addition to the productive capacity of the cells, the sensitivity of some of the cell lines to interferon showed a slight decline[27].

The appreciation of such variation is important when comparing the results of one laboratory working with a particular diploid cell line and another laboratory with either the same or different cell lines. This is particularly relevant when considering the growth of cells on microcarriers where the characteristics of the growth system are likely to accentuate the effect of small differences in the quality of the cells or other components of the system.

Established cell lines. Cell lines which are capable of growth for an indefinite number of passages are included in this category. Such cells are not used for the production of vaccines or other materials for human use. They may be used for the production of veterinary vaccines (FMD vaccine)[22]. During the course of many growth cycles, BHK monolayer cells may change considerably from their original fibroblastic character. This is particularly obvious when the cells are examined in bottle cultures. "Normal" growth is characterised by a "frosted glass" pattern of whorls, while other BHK monolayer cells may have whorls which are only visible under the microscope or cells which are epithelial in form or even cell sheets in which clumps or nodules may be seen. In addition to such physical differences, the cells show marked differences in susceptibility to a range of FMD virus strains[67].

As well as variations in appearance and susceptibility, differences of growth rate or serum dependence may be seen[68]. It is clear that most of these changes are characteristic of cells which have been transformed. From the extensive literature on the characteristics of transformed cells[69–72] it may be possible to define the degree of transformation of the BHK monolayer cells and relate this to the susceptibility of cells to particular strains of FMD virus.

4.3 Surfaces on Which Animal Cells Grow

The processes of cell attachment, flattening out and cell division are all dependent on the relationship between the cell and its supporting substratum or surface. Changes in cell shape during the growth cycle have been vividly pictured in scanning electron-micrographs[73] and the relationship between cell shape and cell division has also been characterised[74].

Materials used for physically supporting the cells during growth. A wide variety of material types have been used to support cell growth (Table 2). Many of the materials are coated with serum, protein or polymers before use. All are used in a medium which contains a complex mixture of macromolecular and polymeric materials. Not only is the detailed chemistry of the surface important but the physical form in which that surface is presented to the cells is influential. For example, single BHK cells would not grow on glass beads smaller than 20 μ in diameter[75] and the choice of the appropriate surface contours of metal plates was found to be useful for the growth of a variety of cells[76].

4.4 The Cell Surface

The functional entity which defines the relationship between a cell and its exterior is a highly complicated multi-layered structure with many different chemical elements and physical components. In addition, the surface is a dynamic structure in which various molecular components may move and, during the course of a cell cycle or during locomotory activity, further changes in the dispositions of the various molecules which make up the cell surface occur. Table 3 summarises a portion of the data on the gross composition of the cell surface.

Table 2. Materials used for supporting surface-dependent cell growth

Material	Treatment	Ref.
Glasses		
□ Window glass	alkali wash	77
	steam	78
□ Borosilicate glass		22
Metals		
□ Steel		76
□ Titanium		76, 8
Palstics		
□ Polystyrene	$KMnO_4$	80
	poly (2-hydroxyethylmetharcylate) (Poly HEMA)	74
	corona discharge	79
	concentrated H_2SO_4	81
	u.v.	81
□ Melanex	± surface roughend (physically)	82, 88
□ Polycarbonate		83, 84
□ Acrylonitrile/vinyl chloride (plus others)		86
□ Teflon (PTFE)		87
Carbohydrate polymers		
□ Cellophane		4
□ Cellulosic sponge	collagen	89
□ DEAE Sephadex A50 (standard)	collodion (cellulose nitrate)	90
	serum	91
	carboyxymethyl cellulose	92
□ DEAE Sephadex A50 (low charge)		93
□ Cellulose acetate-silicone-polycarbonate		95

Basically the molecules associated with the outer plasma membrane of the cell are proteins, glycoproteins, lipids, glycolipids and lipoproteins. Each of these molecular species may carry a net charge as a result of the basic groups of amino-acids and amino-sugars or acidic groups as on amino-acids, sulphated polysaccharides and N-acetylneuraminic acid residues. The overall effect is that cells have a negative charge and migrate to the positive electrode in electrophoresis experiments [156)].

The way in which the various molecules are organised in the plasma membrane has received much attention. It was shown that all the carbohydrate-containing molecules have the carbohydrate portion on the outside of the membrane [102, 103)]. Furthermore, the protein molecules were either nested into one side or the other of the membrane or traversed the whole membrane. The latter molecules are said to be amphipathic in that the part of the molecule within the double layer of lipid molecules is composed of hydrophobic amino-acids while the external portion of the molecule mounting the carbohydrate residues is hydrophilic [104)]. From the calculations in Table 3, it is inferred that most of the lipid molecules in the outer layer of

Table 3. Some components of the fibroblast cell relevant to the way the cell relates to its substratum

Component	Approximate	Ref.
Surface area	1000 μm^2	
Total cell protein		
□ – quiescent	1200 pg	94
□ – growing cells	350 pg	94
Weight of cell membrane (p=1.15, thickness=75 Å)	8.6 pg	98 (by calculation)
Weight of membrane protein	4.3 pg	98[a]
Weight of membrane lipid	3.4 pg	98
Weight of membrane carbohydrate	0.7 pg	98
Weight of carbohydrate in sphingolipid	0.05 pg	98
No. of lipid molecules/protein molecules	70	99
"Cell surface protein" (CSP)[b]		
□ – quiescent cells	25–70 pg	94, 96, 97
□ – growing cells	1–10 pg	94, 96, 97
No. of lipid-bound sialic acid residues	840 x 10^6	100 (by calculation)
No. of neutral sphingolipid molecules	200–300 x 10^6	101, 100 (by calculation)
Total glycolipid molecules	1040–1140 x 10^6	
No. of protein molecules (assuming average molecular weight of 100,000)	26 x 10^6	
Therefore no. of lipid molecules	1820 x 10^6	

[a]Proportions assumed to be as for red blood cell ghosts
[b]Cell-surface protein or CSP or SF or LETS or Band I or Glycoprotein I or Cell Adhesion Factor (CAF), mol. wt. 200,000–250,000 (94, 96, 97)

the membrane also have carbohydrate residues containing the acidic N-acetylneuraminic acid residues.

All the molecules in the membrane are mobile. This can be seen most vividly in the caping phenomenon induced by the interaction of the glycoproteins in the membrane with the plant lectin concanavalin A[105, 106].

In addition to the double-layered membrane described above, some of the proteins which are associated with the membrane are attached to an underlying network of microtubes and microfilaments[108–110] which together comprise a cytoskeleton. Both actin and myosin types of protein molecules are held to be active in such structures[107]. It is thought that many of the changes in cell shape, cell-to-cell adhesion, and cell-to-substratum adhesion occur as a result of changes induced in this cytoskeleton[111].

4.5 Interactions Between the Cell Surface and the Supporting Substratum

The interaction between the cell and its supporting substratum normally occurs in a medium containing serum, thus adding to the complexitiy of the system. In spite of such difficulties recent work has discovered many new features about the association between the cell and its solid support.

(a) A glycoprotein (called the cell-surface protein or CSP) with a molecular weight of 200,000–250,000 Daltons has been implicated in the attachment of the cell to its substratum and to plant lectins (Table 3[95–97]). The protein is produced by the cell and may constitute up to 3% of the total cellular protein[96]. From calculations on the weight of the cell membrane, not all of this protein is present in the membrane and it may be excreted from the cell[94]. This protein will also restore the ability of a transformed cell to spread, adhere to the substratum and exhibit a degree of density-dependent growth inhibition[97].

(b) By using a new chelating agent which sequesters calcium ions specifically, (EGTA) it has been possible to examine in more detail the "footprint" which is the remains of a vesicular structure left behind on both plastic and glass substrata after the cells have been released by the removal of calcium ions[112]. One of the glycoproteins present in this material has been identified as the material described in (a) above. Other proteins include a myosin-like protein, an actin protein and histones (although it is thought that the latter have leaked out from within the cell). In all, some 15 species of protein molecule have been found in the footprint[112].

(c) The implication of bivalent alkali metal ions (Ca^{++}, Mg^{++}) in the attachment of cells to the substratum is well known. It has been reviewed[113] and the inclusion of magnesium acetate in the medium has been shown to help the adhesion of cells to roller bottles[114].

(d) Opinions have differed on the influence of serum in the cell-adhesion process. Some investigators[77, 115] have indicated that serum reduces the attachment of cells to the substratum while others[95] have found the cell-surface protein (CSP), [cf. section (a) above], to be present in the serum. It has also been shown that 10 different serum proteins are associated with the surface substratum after it has been washed in EGTA[112].

(e) It is thought that the removal of calcium ions by EGTA and the action of trypsin on cells which are attached to the substratum causes a disintegration of the cytoskeleton, either locally or generally, thus releasing the cell from the surface[112, 109, 118].

(f) Considering the nature of the substratum, there seem to be two factors which can be identified as having a controlling influence on cell attachment and spreading. The first is the macromolecular organisation of the substratum. Agarose (a polygalactose carbohydrate) is a neutral material which generally does not support the growth of untransformed cells. However, if the gel is allowed to form slowly under conditions where the rate of cooling is carefully controlled, the galactose chains form a double helix which generates a regular fibrillar-like structure in the gel. Such a preparation supports cell growth and adhesion. The same material, cooled rapidly, has a structure where the organisation of the molecules is highly localised; this latter preparation does not support cell growth[116].

The second factor hinges on the charge structure of the substratum. It would seem that cells will adhere to most surfaces which express charged structures of both negative and positive net charges but that neutral materials present the greatest

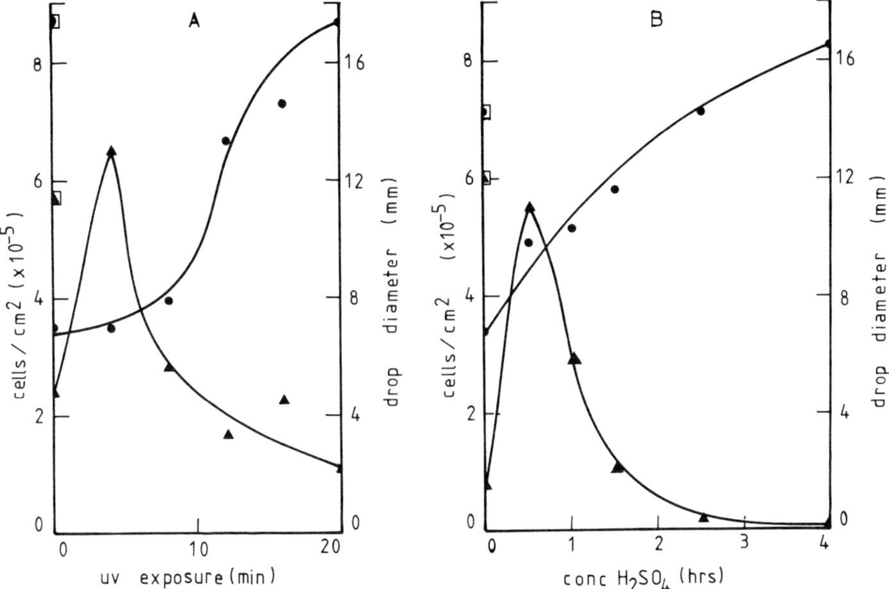

Fig. 2a. The relationship between the time 7 cm "bacterial grade polystyrene Petri dishes" were exposed to u.v. light (at a distance of 10 cm) and (i) the wettability of the plates as determined by drop diameter (•) (ii) the yield of cells per cm^2 after incubation at 37 °C for 96 h (▲). The boxed symbols on the ordinate were obtained with untreated glass Petri dishes

Fig. 2b. The relationship between the time 7 cm "bacterial grade polystyrene Petri dishes" were treated with concentrated sulphuric acid and (i) the wettability of the plate as determined by drop diameter (•) and (ii) the yield of cells per cm^2 after incubation at 37 °C for 96 h (▲). The boxed symbols on the ordinate were obtained with untreated glass Petri dishes

problems[116]. This is borne out by work in the author's laboratory where bacterial grade polystyrene dishes which do not wet easily can be transformed into wettable dishes by oxidising agents ($KMnO_4$), strong acids (H_2SO_4, chromic acid, nitric acid), u.v. light and the corona from an electrical discharge[79]. In two cases an optimum degree of treatment was observed for growing cells (Fig. 2). In a different context it was also shown that electrically neutral Sephadex would not support the growth of cells while the positively charged DEAE Sephadex would[116]. It was also reported that there was an optimum for cell growth with a charge density on the DEAE Sephadex microcarriers of about 2 meq g^{-1} dextran[117].

(g) Poly HEMA [poly (2-hydroxyethylmethacrylate)[74]] is a polymer which, when added in varying thicknesses on the surface of polystyrene dishes, decreases the adhesiveness of cells to the substratum in a controlled manner. The decrease in adhesion results in cells which do not flatten out and in the limiting case may stay rounded. DNA synthesis and division in untransformed cells was also shown to parallel the degree of adhesiveness as flattened cells would divide while rounded ones did not.

This effect did not apply to transformed cells. Tissue-culture grade polystyrene dishes are negatively charged and the neutral hydrophobic hydrogel could act by controlling the intensity of this charge by defining the distance between the source of the charge and the cell. In this way the stimulus to cell flattening may be modulated.

It is clear that the interaction between a cell and its substratum in the presence of serum is a complicated phenomenon. It can probably be broken up into several steps. The first may involve the adsorption of serum proteins on to the substratum[211]. This could be followed by the attachment of cell-surface components to the already adsorbed serum proteins[112]. Such an attachment could take place through the CSP. Then molecules of CSP remote from the initial point of attachment may move into the attachment zone and effect a consolidation of the initial contact. Actin filaments would then bridge the junction between the cellular end of the CSP and the microtubule assembly beneath the cell surface. What does seem certain is that the cell is not in contact with the substratum over the whole of the adjacent cell surface. Electronmicrographs of footpads[112] and whole cells[118] show that, when trypsin and calcium-ion-sequestering agents disturb the cytoskeleton, the cell looses its "grip" and the remaining bond between the cell and the substratum is based on physical attractions or "stick" in localised areas of the cell membrane[118]. Serum is not an obligatory component of this chain of events although it is generally present in most systems which seek to grow untransformed cells in the monolayer mode. Whether it is necessary to insert a molecular bridge between the substratum and the cell will depend on the characteristics of both components and remains an unanswered question.

5 Advantages of Unit Process Systems

Multiple processes involving the performance of the same operation several thousand times are clearly not efficient. While many such operations can be highly automated[119, 120], yet, with the exception of temperature control, little influence can be exerted over the environment within each bottle. When such processes are scaled-up, the buildings, equipment and labour are scaled up commensurately. Thus there is little to be gained from operations on the larger scale. The expected advantages from unit process systems are presented in the Table 4. At present these advantages have not been fully realised in the field of surface-dependent animal cell cultures (cf. [13]). However, by analogy with the history of the development of penicillin production systems, where growth of the mould on trays was rapidly superceded by large-scale deep-tank operations of up to 75,000 l capacity[121], it is not unlikely that a similar scale up of tissue-culture systems would yield similar advantages.

At the time of writing it is not possible to put values on the cost advantages of unit process systems as opposed to multiple process systems. The unit processes under discussion are less than 10 years old and the special buildings and trained operating personnel have not yet emerged. So, in order to discuss the relative merits of the

Table 4. Advantages of unit process systems

1) Enable production at an economic price	
2) Increase profitability by:	
a) Lower capital costs	building equipment
b) Lower operating costs	less labour more efficient use of materials more efficient use of services less contamination
c) Yields per unit cost higher	better control over system larger batches
3) Improved standards of safety and working conditions	

various unit process systems it is necessary to examine in detail the operational characteristics of each system and then by a critical analysis to highlight those systems which have the potential for further development or have the capability of fulfilling a unique need.

6 Operational Features of Systems Designed for the Large-scale Cultivation of Surface-dependent Animal Cells

The operational features of large-scale tissue culture systems may be divided into physical, biological and commercial sections. The physical features of the system consist of the hardware and the manner in which it is operated, while the biological section deals with the kinds of materials which can be grown and produced in the equipment. The final section examines the large-scale production process from a manufacturer's standpoint and presents some of the questions a manufacturer would ask before deciding on which system would be of most potential benefit to his organisation.

6.1 Physical Features

Ability to scale-up. At present, surface-dependent animal cell cultures have not been scaled-up above 100–200 l medium volume. As the requirement for large cultures could be of several thousand litres in size, equivalent to current practice in the field of suspended animal cell cultures[122, 127, 123)], it is clear that the present design of a surface-dependent system should enable scale-up to take place with the minimum modification to the system geometry or mode of operation.

The extensive literature on the scale-up of bacterial and fungal deep-tank cultures[124–126)] attests to the complexity of scale-up operations. An additional problem presented by monolayer tissue culture is the biphasic nature of the system. However, there are monolayer systems which are homogenous, such as the microcarrier cultures and other systems where the fluid component is thoroughly mixed and irrigates the solid phase of the system evenly[78]. Systems made from stacked plates[84] and bundles of hollow fibres[86] did not permit the regular flow of medium across the surface and modifications to these systems were made to obviate this. It remains to be seen whether the redesigned small-scale prototypes are any better than their predecessors. Similarly it is unlikely that spiral-film systems when operating in the vertical mode provide an even distribution of medium.

The scale-up of rotating plate systems also presents problems:

(a) When the plates are increased in diameter, the circumferential speed, which defines the shear effects of the medium on the cells growing on the surface of the disc increases, even if the rate of plate rotation is held constant.

(b) As the plates are mounted on a shaft the bending moment of the shaft increases with the length and weight of plates. (This will cause problems with end bearings or require the installation of additional bearings in the middle of the unit, thus restricting the ease of use.)

(c) As the surface in such units is generally a limiting factor, it is important to take advantage of both sides of the plates. Thus the equipment is rotated through 180° for planting cells on the second side and then rotated a further 90° for the cell growth[128]. As such equipment increases in size the supporting framework becomes massive and the equipment increasingly difficult to handle.

(d) Also, all connections to the unit have to be either broken or flexible for the orientation operation which leads to further complications and potentially dangerous working conditions.

Scale-up on the basis of maintaining a fixed rate of transfer of oxygen from the gaseous to the aqueous phase[127] has not been exploited for monolayer systems. Rather, dissolved oxygen concentration has been held within noncritical limits[24].

In large-scale microcarrier cultures agitation is arranged to be just sufficient to maintain the microcarrier in suspension. It is preferable to use a large impellor roating slowly rather than a small impellor spinning rapidly. In this way the shear forces are minimised and the delicate Sephadex microcarriers are less likely to disintegrate (Fig. 6D).

Finally, systems based on sheets of window glass do not scale-up readily, for, in bigger units, the temperature cycling during sterilisation is often sufficient to break the large glass sheets.

Surface/volume ratio. The importance of this parameter depends on whether the equipment is used for cell generation or product manufacture. For cell growth, providing the surface available is not limiting, the number of cells produced is related to the volume of medium used (Fig. 3). This means that the size of the system may be defined in terms of the volume of medium it contains. This relationship does not hold for virus production where an 80% overall yield of FMD virus may be obtained in 30% of the medium volume used in the growth phase (Fig. 4).

Although the term "contact inhibition" is no longer applied to cell growth[129, 130], the number of cells which will grow on a given surface area is very variable and depends on the type of cell and a variety of factors related to the composition and

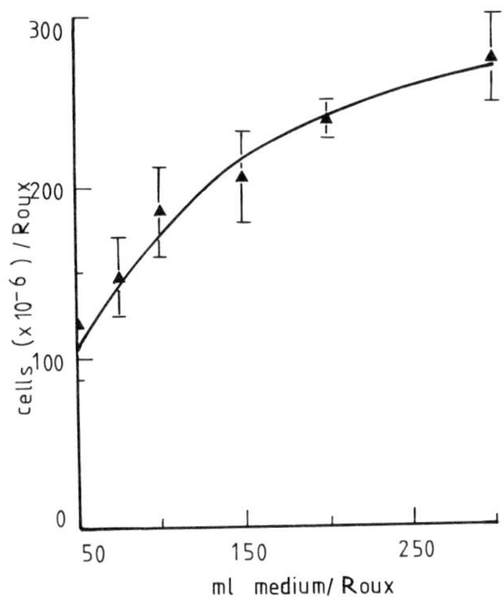

Fig. 3. Cell yield obtained at 96 h from Roux bottles planted with 10 x 10^6 cells in different volumes of medium. The box associated with each point portrays the standard deviation (data from 4 experiments averaged)

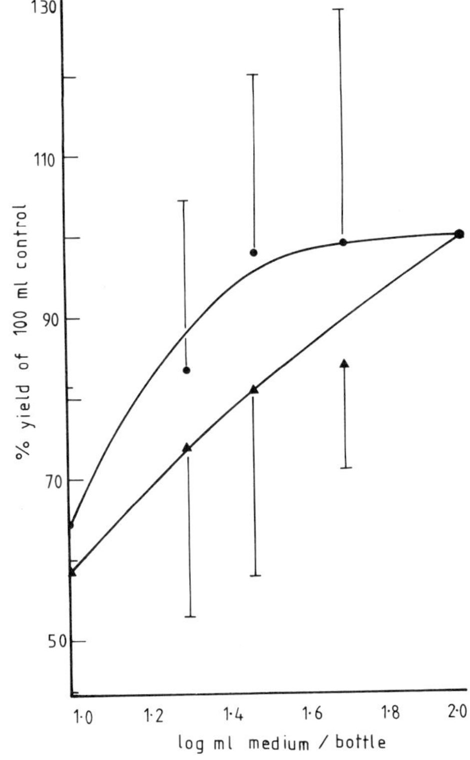

Fig. 4. Percentage yield of FMD virus strain Asia I Iran 1/73 (▲) (5 experiments) and O_1 BFS 1860 (●) (14 experiments) grown in different volumes of medium in roller bottles, compared with a control containing 100 ml medium. The bars associated with each point show the size of the standard deviation

manner of application of the growth medium. Thus cells which are said to be "density-inhibited" may be induced to divide further when stimulated by medium changes[131, 132], additional applications of serum[133], small molecules such as amino-acids, glucose and phosphate[134], steroids[135], pH changes[136], growth factors[137], and polypeptide hormones[138]. Also, cells which were thought to grow in monolayers only have been induced to grow in multiple layers when medium was supplied continuously in a perfusion system[139]. In such a system WI38 cells were induced to form a sheet 5 cells thick.

Whenever the designers of equipment to produce surface-dependent animal cells on the large scale felt that the system was surface-limited, they sought to pack as much surface as possible into a given volume of equipment and then they made sure that all of it (both sides of plates, films, spheres)[140] was made available for the cells. However, in systems where the surface is provided by microcarriers or glass spheres it is possible to confine a large surface area in a relatively small volume. In these cases it is the volume of medium needed to satisfy the amount of surface available which becomes the limiting and hence controlling factor in the determination of the size of the equipment. (For BHK monolayer cells 1 m^2 surface needs spproximately 5 l medium to produce a cell sheet with 6 x 10^5 cells per cm^2.)

The second phase of the operation, the production of a virus or cellular biochemical, is not generally dependent on the volume of medium. It is advantageous in this phase to use as little medium as possible so as to produce a product with the highest concentration. Thus those unit process systems which lack the flexibility of operating at different surface-to-volume ratios will be less advantageous than those which can.

Operational characteristics. When a process is easy to operate there are fewer errors and the likelihood of contamination or malfunction is correspondingly lower. The improved reliability and reproducibility so obtained makes an important contribution to the overall efficiency and profitability of the operation.

Washing. This is the first operation of the production cycle. Equipment which has to be disassembled before washing suffers from the disadvantages (a) as the scale of the operation increases the difficulty of disassembly also increases (b) in the reassembly operation, it is often necessary to assemble specific pieces together in defined orientations, (c) if the assembly is incorrect leakage and contamination occur, (d) the time, labour and space required in the washing area are all important cost factors. For these reasons disposable plastic units[141–143] or units which can be washed in place[78] have advantages.

Sterilisation. The largest item which can be conveniently autoclaved is about 100–200 l. Most units of 100 l and over would be sterilised where they stand. Equipment which changes its orientation during the cell-planting or cell-growing modes of operation has to use either flexible steam hose or couplings which both disconnect and seal. When microcarriers are used, the beads are sterilised outside the unit either by steam (Sephadex-based beads) or irradiation (polystyrene). Such beads may then be fluidised in the medium and transferred into the culture vessel as a suspended solid. Again, disposable plastic systems offer some advantages although the size of such systems is relatively modest.

Fluid transfers. Fluid transfers to and from the culture unit are involved in the following operations (a) planting cells (four operations for some plate systems), (b) refeeding or perfusing cells during the growth phase, (c) washing or treating the cells, (d) addition of medium into which product is liberated, and (e) discharge of product. It is clear that such operations should be easy to do with great reliability. In two special cases it is important to remove as much of the cell-growth medium as possible. The first is during the preparation of vaccine for human use where traces of non-human serum (normally foetal bovine serum) could cause allergic reactions. In this case the general practice is to wash the cell sheet two or more times with phosphate-buffered isotonic saline solutions. The number of washes is determined by the liquid hold-up in the system after the spent growth medium has been discharged. This could be a disadvantage of microcarrier type systems where the volume of retained fluid is relatively high and the time required to resediment the beads and filter off the wash fluids could be significant. The second case is the superinduction of interferon[144, 153]. In this system, translation of induced interferon mRNA is blocked by the inhibitor cycloheximide. This inhibitor and the inducer (normally poly I–C) are then removed and interferon is produced by the translation of the induced mRNA. Also, if this latter operation is done in the presence of actinomycin D (which blocks DNA-dependent RNA synthesis), the time over which the interferon is produced is extended. For such a system to work it is necessary that the inducers and inhibitors should be added and removed from the system with the greatest speed and facility. It is possible that the timing and completeness of such operations is crucial to the successful induction operation.

Sampling. A sample removed from a system should, as nearly as possible, represent the whole system. In the biphasic systems covered by this review this kind of sampling operation is only possible for microcarrier systems. Here, removal of the beads on which the cells grow is a valuable feature of the system. It is discussed further below. With the other systems it is generally possible to obtain a sample of the fluid bathing the cells and, where mixing is efficient, this may be taken as representative of the whole. Mechanisms for the removal of a sample with the minimum risk of contaminating the culture vessel have been reported[127, 145]. The use of an evacuated bottle with an exposed rubber diaphragm which is pierced by a needle attached to a thin pipe whose distal end is in the middle of the vessel is a useful method of sample removal. It has the advantage that, on removing the bottle from the needle, the diaphragm wipes the needle free from drops, which decreases the hazard to the environment from a contagious virus or toxic product.

Infecting. It is a common practice in virus production systems to add the virus to the cells in as concentrated a preparation as possible. This facilitates the adsorption of virus. Those systems which can operate at a higher surface/volume ratio during this phase of the operation have an advantage over the systems with a fixed surface/volume ratio.

Space requirement. Many authors[146–148] have stated that the unit process system they have been considering is highly compact and therefore saves space in the laboratory, incubator or building in comparison to roller bottles or stationary bottles where about 90–95% of the system is taken up by the gas space. This gas space is clearly eliminated in a system where the medium is gassed by blowing air or oxygen into it. However the volume of the system is controlled by the total amount of medium required for growing the cells so that, although the surface available for cell growth may be condensed into a very small volume (as with microcarrier systems), the actual volume required will be similar to other systems which seek to produce the same number of cells.

Robustness and simplicity. In order for a piece of equipment to become part of a production process it must be robust and simple. In routine production most pieces of equipment can be expected to be dropped, bumped or otherwise maltreated. Units which do not require very careful handling are not made of, or contain, glass or other easily fractured materials will survive longer.

Simplicity is also an important feature. In simple equipment, faults are easier to trace and worn or broken parts are readily replaced. The actual operation of the equipment is facilitated, both in a reassembly operation and in running. Furthermore the equipment may be used in areas where technical support staff are not readily available.

Safety. Considerations of safety must include both the safety of the operating personnel to the biological or physical hazards posed by the equipment and also the safety of the environment around the production facility. Delicate vessels made of glass or thin-walled plastics pose a severe risk. Large-scale units made from stainless steel could be protected by (a) double valving, (b) incinerators on outlet gas lines, (c) well-designed sampling equipment (see 'Sampling' above), (d) welded pipework and (e) adequate alarm and monitoring systems to warn of impending failure.

In addition, consideration must be given to the building in which such equipment is housed. While it is outside the scope of this chapter to consider this in detail, much work is in hand on methods of monitoring and designing buildings with carefully controlled air flows between the various sections of the building and between the building and the exterior.

The safety of personnel who operate large-scale equipment is also of paramount importance. Heavy, complicated and intricate equipment generates hazards. Equipment with moving parts and many electrical connections also presents problems.

6.2 Monitoring and Control of Process Parameters

The ability to monitor and control the reactions which occur in the culture systems is one of the major advantages which unit process systems have when compared with multiple processes. The extent to which it is practicable to achieve such control will determine the relative value of the various systems discussed in Sect. 7 below. Of all the systems proposed for monolayer cell cultivation the system based on microcarriers is clearly the most controllable as it is homogenous and approximates to a suspended culture of animal or other microbial cells. Other systems operate with the surface for cell growth in a rigid structure. Within such structures it is possible that irregularities develop. Thus for both the static-plate and the hollow-fibre systems[84, 86] second generation units were developed to overcome such problems. It is likely that other systems based on a fine matrix of particles [149, 150] will also operate at reduced efficiencies. Cells grow across the gaps in a matrix to such an extent that they may form a continous sheet over a 100 mesh stainless steel screen [44] and can often be seen bridging the gap between two closely associated microcarrier particles (Fig. 6C). In situations like this the medium may be completely mixed and therefore homogenous but the irregular and uneven flow of medium, caused by the disorderly blocking up of the pores of the bed or matrix, results in the ragged growth of cells and the development of local pockets where cells might die and degenerate. Similar irregularities exist in titanium plate propagators where the thin metal plates may not be exactly parallel. This causes heavier planting of cells where wide gaps occur and very few cells in the narrow gap areas.

Many systems are made up of two separate vessels, one to hold the surface and some medium and the second to hold the remaining medium [78, 147]. The medium is made to circulate round the system by a pump. If the rate of recirculating the medium over the surface is adequate the difference in the pH, glucose concentration, dissolved oxygen level or other consumable or waste product concentrations between the inlet and outlet ports of the vessel containing the surface will not be great and may be overlooked.

There is little published work on the effect of physical parameters on the growth of monolayer cells in unit process systems. The literature for suspension cultures of animal cells is more extensive [127, 151]. Care must be taken in interpreting such results, as cells which can grow in the absence of a substratum are considered to some degree transformed and will generally exhibit differences in energy metabolism [152] as well as differences in cell-surface features [70].

Temperature. Control of temperature may be obtained by siting the equipment in an enclosure whose air temperature is controlled. In such circumstances, unless very great care is taken, temperature gradients of 1°–2°C may occur across the enclosure, leading to quite large differences in cell yields. Alternative methods of control involve circulating water, maintained at a controlled temperature, around the jacket of a unit process propagator. In such a case either the temperature of the water or the circulating pump output is modulated by temperature differences. This simple system works well for most large-scale units although some difficulty could be experienced constructing vessels which require a special conical geometry in their lower section [155].

pH. Control of the pH of a culture at a single value may not be the most advantageous way of obtaining the maximum yield as the cells may require a different pH value for attachment, spreading and for growth and division. There is little doubt that the size of the net negative charge on the cell surface is dependent on the pH of the medium [156] so that the relationship between the cell and the substratum is likely to be sensitive to the pH of the bulk of the medium.

The measurement of pH is simple in the systems which are well mixed but units whose entire working volume is filled with distributed rigid surface may have difficulty finding space for the probes or a position in the vessel where the medium is moving past the probes sufficiently rapidly so as to represent the bulk of the medium. In such cases an external loop around which medium is propelled by a peristaltic, diaphragm or airlift pump may be used to house the probes. The disadvantages of such a system are (a) increased complexity, (b) problems in sterilisation and (c) difficulty in handling the equipment.

The modern meters for measuring pH are accurate to 0.01–0.02 of a pH unit (expanded scale) and stable. In systems where the rate of mixing is not high (i.e. all systems except the microcarrier case), it is advantageous to control the pH by adjustments of either the percentage of CO_2 in the gas stream or the throughput of air whilst using a bicarbonate buffer in the medium. For well mixed systems additions of acid or alkali may be used.

It should be noted that it is not possible to control pH automatically in systems where the whole unit rotates[157)] unless a very precise mechanism rotates the equipment 360° in one direction and then exactly 360° in the reverse direction[158)].

Dissolved oxygen, redox[160)]. The relationship between these two parameters is that the redox potential of the medium depends on the chemical composition of the medium as well as the oxidation state of the various medium components. The oxidation state of the components also relates to the dissolved oxygen level but further changes in such oxidation states may occur when the dissolved oxygen level is either 0% or 100% saturated, (0–160 mm Hg).

In making such measurements, care should be taken to standardise the sensors[167, 168)]. Values obtained for optimal growth of animal cells in suspension are 75–100 mV (O.R.P.)[161, 162)] or 80 mm Hg (dissolved oxygen)[163)]. However for the growth of monolayer cells on microcarriers, a dissolved oxygen level above 25 mm Hg proved satisfactory[24)]. The amount of air needed to obtain this level may be calculated from an oxygen requirement per cell of $0.5–2 \times 10^{-12}$ ml O_2 s^{-1} [164, 165)].

The propagator unit designed from Teflon bags[166)] is capable of bringing cells into contact with the gaseous phase most readily as the Teflon on which the cells grow is permeable to the physiologically important gases. Other systems rely on the dissolution of gas in the culture medium to provide the cells with the required environment.

Carbon dioxide. There is some evidence to show that the presence of carbon dioxide in the gas phase is necessary for the expression of cell functions[165)]. The physical capability to achieve this is dependent on the design of the equipment in a manner similar to that of the aeration system.

Movement of substratum relative to the medium. This may be achieved by either moving the substratum directly or by keeping the substratum stationary and moving the medium past it. In either case the relative motions of the two phases controls the thickness of the boundary layer covering the cells, which in turn regulates the rates at which waste products are removed and nutrients are taken up by the cells. Most investigators have found, therefore, that optimal cell growth occurs at a particular rotation rate for tubes[157)], plates[171, 8)] or medium circulation rate[170, 22)]. In microcarrier systems the prime consideration is to develop sufficient agitation to maintain the beads in homogenous suspension without either breaking the beads or generating conditions where cell attachment to the beads is impeded[24, 169)].

Rotating plates will generate a variable boundary layer along the radius of the plate as the velocity with which a point on the plate moves through the medium depends on the distance of the point from the axis of rotation. However the plate is generally only half-immersed in the medium at any time which enables nutrient transfer to occur by gaseous diffusion and drainage.

The movement of medium in a perfused system can be effected by an air-lift, peristaltic or diaphragm pump. The discharge from each of these pumps pulsates. Peristaltic pumps are less advantageous in that they depend on alternatively squeezing and relaxing a flexible tube which could split or, if a blockage occurs, become detached from its connections. This may cause the loss of a culture, and, where a hazardous virus is in production, a major security risk is incurred. The use of a gas to move the medium around the system[22, 170)] has some advantages in that there are no mechanical parts to fail and the aeration, circulation and mixing may be effected simultaneously.

Monitoring cell growth. It is important to know the state of development of the cell sheet in order to determine the time for cell recovery (for the preparation of an inoculum) or for generating product. As the traditional method of assessing cell growth is by visual (microscopic) means, many unit process systems are designed so that tubes[172)], plates[173)] and spirals[174)] can be observed directly *in situ*. This method of operation could result in a poor assessment of growth as in some instances only a few peripheral plates or turns of the spiral are visible and they may not be representative of the culture as a whole.

Visual observations on samples of microcarriers extracted from cultures however do give a good indication of the state of the culture. In this respect cells grown on the surface of Sephadex-based microcarriers are more readily observed than cells on polystyrene microcarriers (Fig. 6).

In the absence of direct visual examination it is possible to infer the extent of cell growth by observing variations in parameters which change as a function of cell number. Such parameters could be glucose concentration in the medium[175, 78, 200)] turbidity[166)], pH (colour of phenol red indicator) or the gassing rate where the dissolved oxygen level was controlled and pH adjusted by addition of acid or alkali. In each case correlations established in small-scale propagators should be checked in larger units.

Larger units tend to be built out of stainless steel so that (with the exception of microcarrier systems) it will become increasingly necessary to use indirect methods to monitor the growth of the cells. Automated equipment adapted to this function could prove valuable in many situations.

Monitoring cell breakdown. The growth of many virus strains causes cellular breakdown or other cytopathogenic effects. It is quite common to estimate the time to harvest a virus culture by the visual examination of the cell sheet. While such examinations are possible for microcarrier systems (at all scales of operation) and for transparent culture vessels (see previous section), it is not possible to see the cell sheet in larger units built from stainless steel. Instead, the parameters which may be monitored are (a) enzymes released from infected cells such as isocitric dehydrogenase[175)], glutamic-oxaloacetic transaminase[175)] and lactic dehydrogenase[176)], (b) enzymes which are associated with the virus produced such as neuraminidase, (c) antigens associated with the virus assayed by a complement fixation[177)], radioimmunoassay[178)] or ELISA technique[211)], and (d) observation of physical variables such as pH or dissolved oxygen level which, at a constant rate of aeration, should increase to a maximal value when all the cells have ceased to metabolise [193]. Again, such assays may be effected with automated equipment.

6.3 Biological Features

Growth of required cells. There are many constraints which are applied to a manufacturer which severely restrict the choice of cell type which may be used to generate a particular product (a) the cell must be able to produce the particular virus or biochemical required, (b) such cells should be recognised by the regulatory authorities as an approved source of materials for human use and (c) the cells must flourish in the equipment chosen for their production.

Equipment, therefore, should be able to support the growth of primary cells and human diploid fibroblasts (HDF). Very little difficulty has been reported for the growth of primary cells in any type of unit process equipment. However, the growth of HDF on microcarriers has not been achieved with complete success [24].

Cell harvesting. The ability to recover the cells which have been grown in a unit process system is a requirement for the preparation of a cell inoculum for the next larger scale of operation. This operation can be a key feature and can expose the weakness of particular systems. There are four ways in which the cells may be harvested (a) by the use of a proteolytic enzyme (trypsin is the one generally used although papain, pronase, pancreation [200] and chymotrypsin have been examined), (b) by using a chelating agent, conventionally EDTA although EGTA may also be used (cf. D3 above) (c) a combination of enzymes and chelating agent and (d) physical removal by scraping, wiping or stretching the substratum.

The need to trypsinise cells supported on the substratum within a unit process system has given rise to some novel methodologies: (a) after bathing rotating discs in trypsin, the cells may be discharged by increasing the speed of rotation so that the centrifugal force dislodges the cells from the substratum [9, 175], (b) the harvesting of cells from Sephadex after trypsinisation has led to the development of vibromixer method of agitation [90] or the separation of cells from microcarriers as a result of the shear forces generated by forcing the trypsinised microcarrier through a 1.2 mm bore capillary tube [179], (c) cells may be harvested from glass spheres by vibrating the spheres in the presence of trypsin [179], (d) an alternative method is to use trypsin-EDTA mixtures [170] and bathe the cell sheet in this solution.

Two features are of importance. The first is that if the substratum can be concentrated before trypsinisation then less trypsin will be required and consequently less serum for the neutralisation of the trypsin. Secondly the volume of cell suspension will be smaller and this will require less centrifugation or may even be used as an inoculum directly. While there is no evidence that trypsin penetrates into the cell it does cause a considerable disturbance to the cytoskeleton [109]. There is also evidence that when cells are harvested by using trypsin the amount of cell-surface protein associated with the resuspended cells is considerably reduced when compared to cells which have been removed from the substratum by mechanical means [94]. This could affect the attachment and flattening reactions of the cell in a subsequent growth cycle.

To overcome some of the drastic effects induced by the action of trypsin a variety of mechanical means have been proposed to remove cells from the substrata on which they have been grown. In two cases[88, 181], the substratum (a plastic film) is removed from the culture vessel and scraped or stretched so that the cells are dislodged mechanically whilst in a third case the substratum (titanium plates) is wiped by specially designed blades which are actuated within the culture unit itself[180].

A further possibility which pertains to the microcarrier system is the transference of cells from a bead which is fully covered by cells to a freshly added naked bead. Cells which are in the process of division and are therefore only weakly associated with the substratum are most likely to be dislodged by a marginal increase in the shear forces of the system. Such cells could then attach to fresh beads when the system resumed its original state[92]. While this may be possible for some particular cells, it was found necessary to subject BHK monolayer cells to the action of trypsin for a considerable time (2 h[179]) before the cells could be released by the application of mechanical forces. However, a situation of bead-to-bead transfer has been seen with other BHK monolayer cells in a different laboratory but after several such passages the cells behaved as BHK suspension cells and maintained themselves off the bead surface, indicating that the particular BHK monolayer cells used in these latter experiments were already some way along in the process of transformation[182].

6.4 Commercial Characteristics

Having accepted the argument that unit process systems offer advantages over multiple process systems, the manufacturer is presently faced with a variety of unit process systems to choose from (Sect. 7 below). He would wish to know which systems are capable of reliable operation with his cell/product system at the scale of operations at which he wishes to work. A second question would be, 'Are such systems available "off the shelf" for his "in-house" use without incurring heavy licensing fees?' and, thirdly, 'Would such systems generate problems in the licensing of the final product?' Having selected one (or more) systems the manufacturer has then to test it (them) to determine whether the product yields and quality offer advantages over either his existing multiple process technology or against values quoted in the literature for other systems.

Product quality. The most important determinant of quality is the potency of the product. For most production systems the product potency is controlled by the biological variables such as the particular cell substrate used[27] or the particular virus seed material or strain[67]. After the preliminary screening to select the cell and virus which yield the most product the next step is to obtain the product in a concentrated form.

The advantages of obtaining a low-volume, concentrated product are many: (a) subsequent product handling operations such as purification, subunit preparation and isolation are very much more efficient, (b) the storage space required to hold product during the extensive testing procedures and during the time necessary to assess demand for particular product formulations may be very expensive if the manner of storage is over liquid nitrogen or in the liquid state at 4 °C, (c) the option of combining more components in a particular formulation of limited volume is possible, and (d) there would be less water to remove in a freeze-drying operation.

Such advantages as are listed above are attainable in those systems with a flexible ratio of surface area/volume. The degree to which such a concentration process may be extended depends on the overall yield of the system, for whilst the product may be concentrated it may only be formed to a fraction of the extent of a more dilute system. An example taken from the FMD vaccine system may be seen in Fig. 4.

Product yield. Yield of product per ml of medium, per cm^2 of surface, per man-hour or other variable is insufficient to ascertain the performance of a unit. A manufacturer will normally look at the productivity of a process over a whole year and assess it in relation to the costs of obtaining the material produced. Thus reliable operations free from contamination will weigh heavily in the balance of advantage, whereas the efficient use of medium or surface is less likely to register as an important factor. A further advantage of reliable uncontaminated operation is that a manufacturer will feel confident enough to "put more eggs into his basket" and operate at larger scales with subsequent economies. Such economies are easily seen when it is recognised that a considerable proportion of the production cost of a vaccine (for example) is incurred by the testing procedures. (A protocol for testing a vaccine for human use may involve some 60–70 individual tests and consume 10–20 l of vaccine.)

Following reliability of performance to a given standard, the capital costs of equipment and buildings and the operating costs of consumables, services, labour and overheads have to be included in the economic assessment of a process.

Regulatory agency factors. Regulations have been formulated which govern most aspects of the production of a vaccine for human use[12)] (see also Sect. 3). The conversion of a bottle process to a unit process will have to be accepted by the agency. For most unit process systems (with the exception of the microcarrier system) the isolation of a part of the confluent cell cultures to serve as uninfected controls is difficult. With packed-bed systems it is possible to use a thief device but the microscopic examination of the withdrawn bed matrix could prove difficult. With these and other systems parallel cultures (at a smaller scale of operations) may serve as controls. Also other parameters may serve to demonstrate the normal state of the cell sheet (glucose utilisation rate at low serum concentration, absence of detectable enzyme release from cells, etc.).

Patents. Most of the systems described in the following section have one or more patents associated with them. To date most of these patents have not been challenged in court so that their utility is unknown. What is clear is that the concept of using DEAE Sephadex A50 beads is not patented although there is an American patent application[183)] to cover the use of the "low charge" variety, a situation which is challenged by the manufacturers of Cytodex 1. A German patent exists covering the system of glass spheres held in a jacketed conical vessel[184)]. However, the principle of using glass spheres as a supporting substratum for growing cells was reported 4 years previously[185)] so that only the details of the German operation may be protected.

A wealth of patents cover the commercial propagators such as the Teflon bag unit[166], hollow-fibre units[86] and some of the processes that are possible in units based on titanium plates[180, 76, 34]. The value of such protection depends on the absence of alternative ways of achieving the same or better ends.

In view of possible complications, potential users of a system must be aware of *all* the patents relating to use of that particular system. Whilst many patents have been used in the preparation of this chapter, the author does not claim to have (a) considered every patent relating to a particular system or (b) appreciated all the legal rights of the patentees.

7 Equipment for the Cultivation of Animal Cells in Monolayers

Monolayer cells have been cultivated successfully for many years in stationary bottles and rotating bottles. It is natural that the basic principles inherent in the use of such containers should be maintained when the surface of many such bottles is consolidated into one vessel. Hence many of the systems which have been designed to supercede the multiple-bottle processes have been based either on a "plurality of stacked plates", which mimics the stationary bottle, or on a multiple array of tubes, which imitates the roller bottle approach. Systems which have extended these concepts further have transformed the rigid plate into a flexible film or plastic bag held in a spiral configuration. The more extreme departures from conventional practice are embodied in systems based on hollow fibres, packed beds and microcarriers. The numerous alternative approaches for the development of large-scale unit process systems from multiple bottle processes have been summarised in Table 5. For the purpose of discussion the various systems have been grouped into eight categories.

7.1 Plates

Sheets of glass and plastic (polystyrene or polycarbonate are the plastics generally used although it is possible to use many others) held in a supporting structure can be used in place of a rack of stationary bottles. Such plates may be round[186] or square[147, 142, 84] and the medium is either circulated through the plate matrix by an airlift pump[186] or it is pumped between the plates by a peristaltic pump operating in an external recirculation loop[147, 84]. In the Multitray unit[142], which most closely resembles a stack of stationary Roux or Brockway bottles, the medium is stationary.

The main problem with units made up of a stack of stationary plates, in which the medium is moved by a pump, is that it is difficult to ensure that the medium passes evenly over the surface of each plate in the stack. Thus the use of an air-lift pump and a magnetically driven stirrer in the early units[186] has been superceded by a system where the medium is fed directly into the gap between the plates by using a discharge

Table 5. Unit process systems for the cultivation of animal cells on surfaces

System Type	Substratum material	Variant Form	Substratum movement	Largest size (approximate Surface/vol. (for product) $cm^2\ ml^{-1}$	Surface area (m^2)	Max. vol. medium	Ref.
Plates	Glass	horizontal	stationary	2	23	100	186
	glass	horizontal	stationary	25	10	(50)[a]	198
	glass/polycarbonate	horizontal	stationary	2	0.5	2.5	84
	polystyrene	horizontal	stationary	3.3	0.6	(3)	142
	glass	variable	rotating	7	0.4	(0.6)	187
Discs	Titanium	vertical	rotating	10	7.2	14	188
	glass	vertical	rotating	4	2.5	12.5	173
	polystyrene	vertical	rotating	6	0.6	(1.1)	143
Tubes (rigid)	Glass	multiple	rotating	>30	1.3	(5.6)	157
	glass	multiple	rotating	15	5	6.5	189
	glass	multiple	rotating	>30	11.4	5.7	172
	glass	multiple	roating	>30	11.4	5.7	172
(flexible)	Teflon	spiral	rotating	?	20	(100)	158
Films	Melanex	spiral	stationary	2	4	20	174
	Melanex	spiral	stationary	5.3	0.85	1.6	141
	Melanex	spiral	stationary	3.4	0.63	1.8	191
	polycarbonate	spiral	rotating	4	0.15	(0.75)	83
Hollow fibres	Organic	fibre	stationary	2	0.005	(0.025)	85
	organic	fibre	stationary	27	0.16	(0.8)	86
Packed beds	Glass	spheres	stationary	12	3.75	(18.75)	155
	glass	spheres	stationary	13	20	100	78
	organic	jacks	rotating	–	–	2(?)	192
	glass	matrix	stationary	>30	(10)	50	149
	kieselgur	matrix	stationary	>30	(200)	1000	193
Microcarriers	DEAE Sephadex A50	bead	suspended	>30	(30)	150	24
	DEAE Sephadex A50 L.C.[b]	bead	suspended	>30	(0.6)	3	93
	Biosilon	bead	suspended	>30	0.005	0.025	194
	glass	bead	suspended	>30	(0.1)	0.5	195

[a]Numbers in brackets calculated on the basis of 5 l of medium per 1 m^2 surface
[b]L.C. low charge

manifold or diffuser[147, 84]. The difficulties of achieving an even flow using this technique will increase as the scale of the unit grows.

Whilst the early systems[186] operated at a fixed surface/volume ratio once the system was operational, the later designs[147] are more flexible as the bulk of the medium is contained in an external reservoir. Instruments to monitor and control process parameters may be placed in the recirculation pipework. However, in order to achieve this concentration of surface the plates have to be close together (<1 mm), adding to the difficulties of distributing the medium evenly.

Problems resulting from the use of window glass, which breaks during sterilisation, may be solved by using polycarbonate or Pyrex, but the equipment necessary to maintain the configuration of the plates is complicated and is unlikely to wear well in routine production operations.

Some of the equipment is protected by patents[84] although this equipment seems to be remarkably similar in its overall operating characteristics to the system described in Ref.[147] which was published in the same year as the patent[84] was filed.

With the Multitray unit, where the medium is stationary, different problems are found (a) as 75% of its volume is gas space it is not very compact, (b) although the recommended limit to the minimum amount of medium which should be applied to the trays gives a surface area/volume ratio of 3.3 cm^2 ml^{-1} (Table 5) it is possible to work at higher ratios if the unit is rocked on special equipment, (c) the units are bulky and are best operated in specially constructed batteries of 4 Multitray modules per battery[196], (d) the cost of the 30-bottle equivalent Multitray package is about the same or slightly less than the equivalent number of plastic bottles, (e) a patent covers the manufacture of the Multitray[197]. There are, however, compensating advantages, (a) the unit is disposable and reaches the user in a clean, sterile, ready-to-function state (b) the number of handling operations is small compared to that involved in the equivalent number of bottles (c) for laboratories which produce a few large batches of cells and cell products in a year, this method, lying between a multiple and a unit process, could be of value. As the Multidish system is relatively new difficulties have been reported in obtaining supplies.

7.2 Discs

The disc system is built up by threading a number of round plates on to a shaft and constructing a vessel in which the shaft and discs may be rotated. Such units operate with the discs in the horizontal plane during cell planting and with the discs rotating in the vertical plane, generally with the unit half full of medium, during the cell growth phase[175]. Two methods are used to obtain cells on both sides of the discs (a) half of the cells are allowed to attach to one side of the disc then the medium is discharged from the propagator, the second half of the cells is added to the discharged medium and the cell suspension is pumped back into the unit which is turned through 180° before incubation overnight, (b) a second method involves mounting the unit in a framework and rotating it bodily with a full charge of cells for several hours at 1 revolution/10 min[128]. Both methods are awkward and are difficult to scale up.

As the surface in the disc units is made to rotate, considerable mechanical problems are encountered. One unit[173] uses a shaft which penetrates the end wall of the vessel and thus constitutes a contamination hazard. Also as the units get larger it becomes more difficult to use a magnetic couple to rotate a heavy stack of discs and wear on the supporting bearings increases the torque necessary to induce rotation. Further problems result from the narrow gap between the plates generating considerable surface-tension forces as the plates rotate out of the liquid[140]. The variation in the surface area/volume ratio is limited as the units have to be at least half full during each stage of the production cycle. In addition the vessel is normally completely filled with the discs so that all monitoring and control operations have to be done on an external loop.

The advantages of using glass or titanium are evident in that products suitable for human use may be made in such units[13]. Whilst numerous patents have been taken out on the application of this equipment[175, 128, 76, 180, 171] the prior art established in 1967[8] could perhaps render the equipment available to other interested parties.

A polystyrene unit based on similar principles[143] is available. It is relatively small in size and could function as a single unit replacing several roller bottles. It is disposable and is easy to use although the costs for a large-scale operation would be considerable.

7.3 Rigid Tubes

An array of rigid tubes held between end-plates which have been specially designed so that all the tubes may be filled, emptied and sterilised as a unit (or bank) have carried forward the principles of the roller bottle into a unit process system. The mechanical requirements have led to the design of complex and unique end-plates and manifolds[157, 172]. This may be circumvented by fixing all the tubes between the end-plates and placing the whole assembly inside a second vessel. Arrangements are then made to rotate the tube array relative to the outer container[189].

Systems of the former design will scale up with little change in the basic geometry if the diameter of the tubes is kept constant. Medium in a tube at the periphery of an array will move around the tube at the same speed as it would in the centre of the array. However, the complexity of the equipment with the concomitant difficulty of disassembly, washing, reassembly and leak testing makes this equipment clumsy to use. Larger-scale units mounted in a massive framework[172] and washed and sterilised in place are difficult to run leak-free as the junction of each tube with the end-plates and the joints of each valve present a potential area for leaks to develop, a problem generally accentuated by repeated cycles of heating and cooling.

The latter system[189] operates like a disc unit. Here as the diameter of the rotating assembly increases the circumferential velocity will also increase; furthermore, as the unit has to be at least half full for all stages of operation, a degree of flexibility is lost.

Although the units described in the literature are made from glass and therefore cause little concern to regulatory authorities, they do constitute a hazard when a dangerous material is produced in them. They would probably work as well if constructed from metal.

A further variant of this system is represented by a roller bottle which encases a number of concentric tubes[190]. This condensation of nine roller bottles into one unit is a half-way stage to a unit process system and may have applications in unique circumstances.

7.4 Flexible Tubes

Having shown that cells would grow on chemically etched FEP-Teflon (Dupont) (a fluor-ethylene-propylene copolymer[181]) which is permeable to oxygen and carbon dioxide) a unit process system was developed[158]) to grow large quantities of cells on the inner walls of a Teflon tube formed into a spiral in such a way that medium could be pumped through the lumen of the tube while the outside of the tube was available for gas exchange. In this way pH, pO_2, and pCO_2 could be monitored and controlled. The equipment designed for this purpose is complex and costly. Each of 20 tubes may be perfused with medium individually (or in series?) while the reels supporting the spiral are rotated backwards and forwards every hour. The possibilities of tube rupture and mechanical breakdown are considerable. A further disadvantage of the system is that all the tubes are controlled as one unit when they may be functioning individually. This development, protected by patents[166, 199]), presents interesting possibilities for examining the relationship between cell growth and the composition of the gas phase in contact with the cells but the complexity and cost of the equipment could exclude it from a routine production department.

7.5 Spirals

Systems based on holding a film of material [Melanex[174]) plain or physically roughened[191]) (polyester) or polycarbonate[83])] in a spiral configuration inside a cylindrical container are included in this category. During the planting operation the cells are added to the unit in a suspension which completely fills the vessel. The filled vessel is then rotated about its long axis overnight to allow the cells to stick to the film. For cell growth the unit is changed to an upright position, when air is passed through an air-lift pump to circulate the medium[146]).

The surface area/volume ratio in such units is not very flexible. The commercially available disposable system[141]) has too high a surface/volume ratio for the growth of cells so that the medium has to be replaced[200]). For the production of virus there is no concentration effect if the units are maintained in the upright position and if they are rotated in a horizontal position the maximum concentration factor is two.

The scale-up of the system presents problems. In the smaller units the even distribution of the medium is not critical but the inherent weakness of getting medium within the space between the spiral elements is accentuated on the large scale. Also the construction of a large spiral is arduous as it becomes increasingly difficult to maintain the gap between adjacent layers. The larger units are also troublesome and heavy to handle and would require complex equipment for the different orientation steps required. In addition the opportunity to apply monitoring and control systems is limited. Whilst it is possible to remove the spiral and release the cell sheet mechanically[88]), when the cells are removed *in situ*, by the use of an enzyme, a large volume of fluid results and low yields are reported owing to the length of time required for the trypsinisation operation[200]).

Such units have been used to advantage for small research operations[146] as they conserve space and, on a modest scale, can be handled with facility on the laboratory bench. The ready availability and ease of use of such units is also an asset.

7.6 Hollow Fibres

The growth of cells on the surface of hollow fibres is an attempt to recreate the kind of environment which cells experience within the body. Considerable progress has been made in this area although the scale of operation has been comparatively small (Table 5). Cells may be grown on the inside or on the outside of the fibres and in either situation they may be induced to react with a specific chemical, such as unconjugated bilirubin converting it to its conjugated derivative. In this form the molecule may cross the membrane and thus become separated from the unreacted material. Numerous other possibilities exist for transformations of this sort[41]. The production of a "rapid harvest" leukemia virus from cells grown on hollow fibres has also been reported[193].

It is unlikely that such systems will be used on the large scale. Most of the hollow fibres are sterilised chemically[86, 40], a time-consuming process which could introduce difficulties in operating technique. In large-scale units there would be doubt that the sterilising gas or fluid had penetrated to every part of the equipment and had permeated particles which had not been removed in the washing process. The main problem is that of homogeneity. Cells grow in clumps across the lumen of the fibres or in masses between the fibres thus preventing the even flow of medium and gas[86]. Attempts to overcome this by making thin pads of fibres[86] may work on the small scale, but the engineering of larger units would lead to the development of a complex structure which is unlikely to be a practical industrial implement. There may however be specific reactions which only living immobilized cells can do, so that a bioreactor may be developed for such applications. In addition, hollow fibres could provide a system which is compatible with human blood and in favourable circumstances could act as an external artificial organ. Such applications would be in the biomedical field rather than in the area of the production of viral prophylactics or antiviral agents.

Much of the information presented has been derived from patents[86, 40, 41] so that others wishing to enter the field should be appraised of the current situation.

7.7 Packed Beds

A packed bed consists of a mass of randomly oriented particulate material held in a container. The choice of material, size and shape of such particles is critical. The ideal bed is one in which, when medium or foam[170] is made to flow over it, each particle is irrigated to the same degree. This results in an homogenous environment for cell growth and allows the proper functioning of control systems, thus facilitating scale up. Elements used in packed beds have been jacks (made from plastics such as methylmethacrylate or polystyrene)[192], diatomaceous earth[193], a glass matrix[149] or 3 mm glass spheres[155, 22, 78]. In each system, with the exception of the plastic jacks, the beds have been stationary while the medium has been circulated over them.

The system of plastic jacks, immobilized in a rotating roller bottle and fed by a perfusion system[192], is a system of limited scale. There is little data on the use of such a unit, yet it would seem that for a laboratory application it could remove some of the risk of opening and closing a number of bottles. In principle, the method could be scaled-up but engineering problems as well as the difficulty of retaining the conditions found effective on the small scale could limit the systems' exploitation.

Diatomaceous earth is commonly used as a filter aid to remove cells from media. There are many grades of different porosity and the application cited[193] describes the growth of BHK monolayer cells "deposited in the carrier bed" as determined by the disappearance of glucose from the recirculating medium. However, no evidence is presented to show that the cells grew on the surface of the particle, so that the particular monolayer cells used may have been transformed in some degree and able to grow independently of surface attachment. In such a system, in which the cells are basically filtered out, the growth of the cells would tend to occlude the spaces between particles and thus cause the circulating medium to seek open channels for easy passage with a resulting uneven growth performance.

A "super culture" apparatus has been described[149] in which cells are grown on a "glass matrix" which has been "layered 1 cm thick on porcelain plates" yielding very large masses of cells. Monolayer cells are highly adept at growing across gaps in a surface structure (see below), and it is likely that the concentrated disposition of the glass surface would encourage such growth, thus leading to irregular flows of medium and eventual blockage of the system altogether. So far the claims of the manufacturers have not been supported by parallel reports in the scientific literature.

In both the kieselguhr and glass-matrix systems the problems of removing cells from the bed has not been dealt with. This omission severely limits the scale of operations at which these units may be used.

For those systems where care has been taken to choose a supporting substratum which will permit medium to flow through the packed bed with ease, the gaps between the bed elements are large (>2 mm) and the bed element selected was the 3 mm diameter glass sphere. Three groups of investigators have adopted such spheres[155, 78, 170] and current work at the 100 l, or larger, scales is in hand. Such systems can be scaled up successfully[78] and cells may be harvested from glass spheres for use as inocula for larger-scale operations[179]. Additional advantages of such systems are (a) the substratum is glass, and therefore is not a matter of concern for a regulatory agency, (b) virtually all cell lines (including human diploid fibroblasts) grow readily and reproducibly on the glass substratum (c) medium can be removed and the cells washed or chemically treated easily and rapidly as there is no sedimentation time involved and little liquid hold-up. (d) As the idea of growing cells on glass spheres was reported in 1970[185] the system, in principle, may not be protected by patents (cf. however [184]), (e) such systems are not available as standard items but are relatively easy to specify and build by process engineering groups (Fig. 5), (f) the systems may be monitored and controlled with ease[202], (g) when the system is scaled up the rate at which liquid is pumped around the recirculation loop does not increase commensurately. Rather,

Fig. 5. Propagators based on packed beds of glass spheres. Medium capacity for L → R 0.1, 1, 10, and 100 l

that amount of liquid is moved which maintains a constant *linear* velocity as it passes through the substrate bed.

The main criticism of such systems is that the cells may not be seen directly. However, Sect. 7.1 included a variety of methods whereby the state of the cells may be determined by indirect means, and regulatory agencies, who generally require that the cells used for the production of a virus vaccine should be available for observation in the uninfected state for some time after the production virus has been concluded, could be satisfied by the inspection of control cultures run in parallel with the larger unit but at a smaller scale of operation.

7.8 Microcarriers

A microcarrier may be defined as a particle of between 100–500 μm diameter which can be held in homogeneous suspension in a stirred-tank reactor. Other properties of microcarriers that would be advantageous are (a) a density of about 1.03–1.05 g cm^3 so that beads may be kept in suspension by the application of least energy and that when the stirrer is switched off the beads will sediment rapidly and completely, (b) a clear optical appearance so that cells may be easily seen as they grow on the surface of the bead (c) there should be no doubt that materials toxic for cells or dangerous for humans and animals are not associated with the particles. The three materials which have been proposed for microcarriers are glass, Biosilon and derivatives of a cross-linked dextran, Sephadex.

Of the three materials proposed for microcarriers, glass in the form of Spherocil [195] offers the least potential. The density of the glass is over 1.5 g cm^3, although the dry bulk powder could have a lower density by virtue of intra-particle space as well as pores of 100 Å within the particle. Such a material is difficult to keep in suspension[90]. Attempts by the author to grow chick or duck embryo fibroblasts on such material were unsuccessful.

The use of a plastic material such as Biosilon [194] in the form of microcarriers has most of the desirable features referred to above. In preliminary experiments the author had no difficulty in growing BHK monolayer cells on this material although it remains to be seen how the many varieties of human diploid fibroblasts perform. Such micro-beads may be observed under the microscope but the growing cells are not as visible on this carrier as on the Sephadex one (Fig. 6E). As this material has only just become available products made from cells grown on it have not yet been licensed for use and the price of the materials has not been established, although the author understands that it is unlikely to be very different from plastic bottles when costed on a surface area basis.

Sephadex. The use of DEAE-Sephadex A50 beads was initiated in 1967 by van Wezel[9]. This pioneering investigation opened up an area which, over the last few years, has expanded into an active and vital field of activity. The main problem with this system was thought by van Wezel to be the "toxicity" of the beads when used at superoptimal concentrations[203]. This was counteracted by coating the beads with nitrocellulose (Collodion)[90] or serum[90, 22]. However, Horng et al.[204] held that the toxicity was likely to be due to the adsorption of "certain growth-enhancing factors" from the medium. In these experiments the beads were not deliberately precoated with serum but were held in phosphate buffer prior to suspension in medium containing 10% heat-inactivated calf serum. In other experiments, however [169], when the beads had been precoated with serum a similar decrease in cell growth was observed when the bead concentration was elevated. At that time the effect was thought to be similar to that observed by Rein et al. [59] where chick embryo cells grew in Petri dishes only if they were planted at a high enough concentration per unit surface area (cf. Ref.[5]). A further possibility is that when the ratio of cells to beads is optimal all the beads become partially covered with cells [about 10–20% confluent (Fig. 6A)]. This makes it less probable for a cell-free bead to collide with a cell which is already associated with a bead. In the event that a second bead becomes attached to a cell already attached to another bead the subsequent motions of the two larger bodies may tear the cell apart. That such collisions do occur can be seen when small beads are seen in association with larger beads with a dense mat of cells formed between them (Fig. 6).

A different approach was taken by Levine an his coworkers[92]. In 1975 they also thought that the problem of toxicity was due to the removal of components from the medium. They decided to decrease the capability of the bead to remove such com-

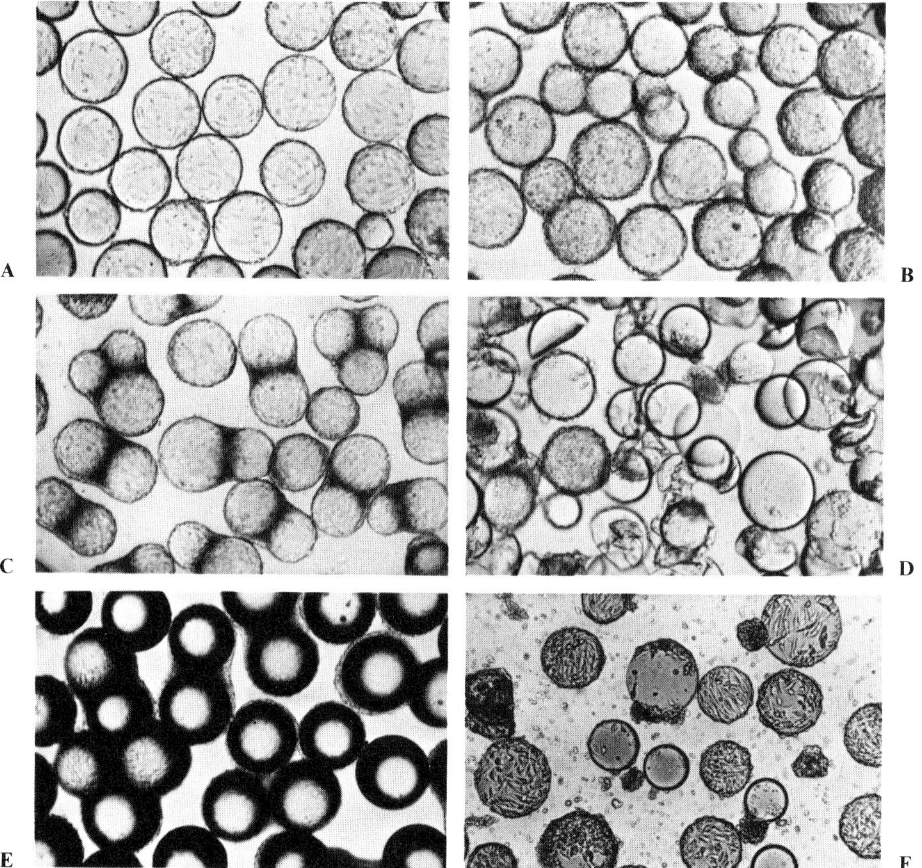

Fig. 6. BHK monolayer cells growing on microcarriers
A Serum-treated DEAE-Sephadex A50 system at 24 h
B Serum-treated DEAE-Sephadex A50 system at 96 h
C Cytodex 1 (low-charge Sephadex) system at 96 h
D Broken beads due to incorrect positioning of the stirrer
E Biosilon beads with BHK monolayer cells
F Nodule formation in medium containing half the normal serum concentration (5%; beads, serum-treated DEAE-Sephadex A50)

ponents by preneutralising some of the positively charged sites on the bead surface with a polyanion (carboxymethyl cellulose). Using this material they were successful in growing a variety of cell types. They developed this approach further by making a DEAE-Sephadex bead with about 1/3 of the positively charged sites[93)] of the regular DEAE-Sephadex A50 commercially available[201)]. This was shown to be an improvement. Later Levine[117)] reported that for the HEL 299 (a HDF) a charge density of about 2 meq g^{-1} dextran was optimal. Below 1 meq g^{-1} cells did not grow and nodule formation was observed. It is of interest to note that when BHK monolayer cells are grown at a low serum concentration on serum-coated DEAE-Sephadex A50 beads, nodule formation is also observed (Fig. 6F). Higher concentrations of serum in the growth medium led to the even growth of cells over the whole surface of the bead[205] (Fig. 6B).

One of the features claimed for the low-charge beads is that they would support the growth of diploid cells and, in particular, those cells which could easily be licensed for the production of interferon[148]. However, reports from various workers, who have used beads prepared according to the procedure of Levine[206] or have obtained low-charge beads under the name of Cytodex 1[201], indicate that the good results which can be obtained with HDF cells are not obtained consistently or reliably.

Consistent performance is a prerequisite of any process which is a contender for large-scale industrial use. The DEAE-Sephadex A50 system used in BHK monolayer cells has also suffered from inconsistent performance[207]. In contrast such erratic performance has not been observed in the primary green monkey kidney cell system as used by van Wezel as it is capable of reliable operations at the 150 l scale[24]. Also Meignier[208] has scaled up a pig kidney cell line system to 130 l using DEAE-Sephadex A50 with 15 mg of carboxymethylcellulose/g dry Sephadex and a bead concentration of 1 g l^{-1}. Cells grown in this system have been used successfully for the production of FMD vaccine.

In a variety of systems it has been shown that it is possible to produce cells at a high concentration in the reaction vessel. In the case of Levine et al.[92] this was achieved by refeeding his system which contained the low-charge Sephadex at a concentration of 5 g l^{-1}. In this way cell concentrations of HDF of 4 x 10^6 cells per ml were obtained and about the same concentration for chick embryo cells[92]. Such concentrations may also be obtained using 2 g l^{-1} standard Sephadex when the cell inoculum is raised to 2 x 10^5 cells per ml[204]. Furthermore the production of BHK monolayer cells at concentrations of 2–3 x 10^6 cells per ml can be achieved when using 0.5 g Sephadex l^{-1} [205] or 3 g l^{-1} [117]. There would therefore seem to be little, if any, increase in the capacity of the low-charge Sephadex to produce cells at a higher concentration compared to alternative ways of using the standard DEAE-Sephadex A50.

The factors involved in the animal cell-Sephadex interaction are clearly complex (cf. Sect. 4.3) and it is unlikely that charge density alone is the prime determinant of the behaviour of the system. Variations in cell quality, serum quality and in the relative concentrations of all the system components (cells, serum, beads, etc.) interact in a complex way to determine the final outcome.

To date, inactivated poliovaccine grown in green monkey kidney cells attached to DEAE-Sephadex A50 microcarriers has been licensed[209] in the Netherlands and the patent position governing the use of the low-charge beads is in process of resolution.

8 Conclusions

It is clear that no one system of growing animal cells in monolayers on the large scale has yet been perfected (cf.[210]). At present much remains to be done to make the microcarrier system both more reliable and versatile and, whilst systems based on glass spheres may be developed to the large commercial scale, much background work on methods of monitoring cell and product formation is necessary.

Also while the microcarrier system could be used to advantage in the production of virus vaccines (polio, FMD, rabies) the ease of washing and chemical treatment of the glass sphere system, coupled with the lack of novelty in the substratum composition, could make the latter system a preferred solution in the interferon production operation.

Other systems, too, may have advantages in particular applications. Hollow fibres could prove more useful in processes which require a particular separation of feedstock and product, while various laboratories which require a large batch of cells relatively infrequently may use disposable units[141–143]. Also plate systems, both horizontal and vertical, could continue to be used in those situations where they are presently in the process of development, whilst tube systems have a limited life in current production operations. Both of these latter systems are awkward to use and do not scale-up readily. Spiral films, too, may have reached the zenith of their development and are unlikely to be used for large-scale industrial production systems.

The developments described above have been crowded into the past ten active years. In the future, further effort will be applied to understanding in more detail the relationship between a cell and the substratum it needs to grow on. By using such information the different kinds of equipment which have been designed and built to operate as unit process systems on the large-scale will become successfully and reliably productive. A result of such productivity is that it will be possible to generate cheaper, safer, easier-to-use products of high potency and wide range of applicability.

9 References

1. Wilmer, E.N.: Tissue culture. Methuen's monographs on biological subjects. London: Methuen & Co. 1935
2. Paul, J.: In: Growth, nutrition, and metabolism of cells in culture. Rothblat, G.H., Cristofolo, V.J. (eds.), Vol. 1, pp. 1. New York: Academic Press 1972
3. Weymouth, C.: In; Growth, nutrition, and metabolism of cells in culture. Rothblat, G.H., Cristofolo, V.J. (eds.), Vol. I, p. 11. New York: Academic Press 1972
4. Earle, W.R., Schilling, E.L., Shannon, J.E.: J. Nat. Cancer Inst. *12*, 179 (1951)
5. Earle, W.R., Bryant, J.C., Schilling, E.L.: Ann. N.Y. Acad. Sci. *58*, 1000 (1953/54)
6. Coulson, J.M., Richardson, J.F.: Chemical engineering, Vol. II, p. 406. Oxford: Pergamon Press 1962
7. McCoy, T.A., Whittle, W., Conway, E.: Proc. Soc. Exp. Biol. Med. *109*, 235 (1962)
8. Molin, O., Hedén, C.G.: Prog. immunobiol. standard. *3*, p. 106. Basel: S. Karger 1969
9. van Wezel, A.L.: Nature (Lond.) *216*, 64 (1967)
10. Mackowiak, C. (Chairman): Developments in biological standardization *35*, p. 482. Basel: S. Karger 1977
11. Document HDC-2 available from the National Institute for Biological Standards and Control, Holly Hill, London, U.K. See also International Assn. of Microbiol. Society Minutes of the 7th Meeting of the Committee on Cell-Cultures, p. 54, 1970

12. Potash, L.: In: Methods in virology. Maramorosch, K., Koprowski, H. (eds.), Vol. 4, p. 372. London: Academic Press 1967
13. McAleer, W.J., Buynack, E.G., Weibel, R.E., Villaregos, V.M., Scattergood, E.M., Wasmuth, E.H., McLean, A.A., Hilleman, M.R.: J. Biol. Standard. *3*, 381 (1975)
14. U.S. Dept. Health Regulations for the manufacture of Biological Products Dept. H.E.W. Title 42, Part 73, Publication Number 71–161. Revised 1 June 1971
15. Drozdor, S., Cockburn, W.C.: In: Int. Conference on the Application of Vaccines against Viral, Rickettsial, and Bacterial Diseases in Man. Scientific Publication no. 226, p. 163. Washington D.C.: PAHO, WHO 1971
16. Miller, D.L.: In: Int. Conference on the Application of Vaccines against Viral, Rickettsial, and Bacterial Diseases of Man. Scientific Publication no. 226, p. 190. Washington D.C.: PAHO, WHO 1971
17. van Wezel, A.: Personal Communication
18. Johnson, M.D.: In: Developments in Biological Standardisation, 42, p. 189. Basel: S. Karger 1979
19. Hayflick, L., Moorhead, P.S.: Exp. Cell. Res. *25*, 585 (1961)
20. Jacobs, J.P., Jones, C.M., Baille, J.P.: Nature (Lond.) *227*, 168 (1970)
21. Nichols, W.W., Murphy, D.G., Cristofolo, V.J., Toji, L.H., Greene, A.E., Dwight, S.A.: Science *196*, 60 (1977)
22. Spier, R.E., Whiteside, J.P.W.: Biotechnol. Bioeng. *18*, 639 (1976)
23. Salk, J., Salk, D.: Science *195*, 834 (1977)
24. van Wezel, A.L., van der Velden-de Groot, A.M.: Proc. Biochem. *13*, 6 (1978)
25. A.F.G., British Patent no. 1, 492, 930 (1974)
26. Scott, G.M., Cartwright, T., Leu, D.G., Dicker, D.: J. Biol. Standard. *6*, 73 (1978)
27. Horoszewicz, J.S., Leong, S.S., Ito, M., Di Bernadino, L., Carter, W.A.: Infect. Immun. *19*, 720 (1978)
28. Buynack, E.B., Hilleman, M.R., Weibel, R.E., Stokes, J.: J. Amer. Med. Assoc. *204*, 103 (1968)
29. Hilleman, M.R., Buynack, E.B., Weibel, R.E., Stokes, J., Whitman, J.E., Leagus, M.B.: J. Amer. Med. Assoc. *206*, 587 (1968)
30. Buynack, E.B., Hilleman, M.R.: Proc. Soc. Exp. Biol. Med. *123*, 768 (1966)
31. Tashjian, A.H.: Biotechnol. Bioeng. *11*, 109 (1969)
32. Shah, U., Jaswal, G.S., Mansharamoni, H.J., Plotkin, S.A., Wiktor, T.J.: Brit. Med. J. 997 (1976)
33. Crick, J.: Postgrad. Med. J. *49*, 551 (1973)
34. McAleer, W.J., Spier, R.E., Posch, K.L., Baugh, C.L.: U.S. Patent no. 3965258 (1976)
35. Mowat, G.N., Garland, A.J.M., Spier, R.E.: Vet. Rec. *102*, 190 (1978)
36. Sabin, A.B.: Brit. Med. J. *1*, 663 (1959)
37. Perlan, D.: Science *160*, 42 (1968)
38. Kohler, G., Milstein, C.: Nature (Lond.) *256*, 495 (1975)
39. Houck, J.C. (Ed.) "Chalones". New York: North Holland 1976
40. Cellco Inc. British Patent 1,395,291 (1975)
41. The Community Blood Council of Greater New York, Inc. British Patent 1,491,261 (1977)
42. Hull, R.N., Huseby, R.M.: Eli Lilly Inc. British Patent 1443189 (1973)
43. Barlow, G.H., Lazer, L.: Thromb. Res. *1*, 201 (1972)
44. Litwin, J.: Proceedings of the First Meeting of the European Socity for Animal Cell Technology, Amsterdam 1976, p. 9. Published by Rijks Instituut voor de Volksgezondheid, Bilthoven, Netherlands
45. Jacobs, J.P.: Nature (Lond.) *210*, 100 (1966)
46. Bolt, K., Clarke, J., Spier, R.E.: In: Developments in Biological Standardisation 42, p. 47. Basel: S. Karger 1979
47. Goodheart, C.R., Casto, B.C., Zwiers, A., Regnier, P.R.: App. Microbiol. *26*, 525 (1973)

48. Boone, C.W., Mantel, T.D., Caruso, E., Kazon, E., Stevenson, R.E.: In Vitro *7*, 174 (1972)
49. Temin, H.M., Pierson, R.W., Dulak, N.C. In: Growth, nutrition, and metabolism of cells in cultures. Rothblat, G.H., Cristofolo, V.J. (eds.), Vol. 1, p. 49. London: Academic Press 1972
50. Litwin, J.: In: Developments in Biological Standardisation, 42, p. 37. Basel: S. Karger 1979
51. Horodniceanu, F.: Proceedings of the First Meeting of the European Society for Animal Cell Technology. Amsterdam 1976. Published by Rijks Instituut Voor de Volksgezondheid, Bilthoven, Netherlands
52. Nuttall, P.A., Luther, P.D., Stott, E.J.: Nature (Lond.) *266*, 835 (1977)
53. Fogh, J., Holmgren, N.B., Ludovici, P.P.: In Vitro *7*, 26 (1971)
54. Keay, L.: Biotechnol. Bioeng. *18*, 363 (1976)
55. Guskey, L.E., Jenkin, H.M.: Proc. Soc. Exp. Biol. Med. *51*, 221 (1976)
56. Tomei, L.D., Issel, C.J.: Biotechnol. Bioeng. *17*, 765 (1975)
57. McKeehan, W.L., Genereux, D.P., Ham, R.G.: Biochem. Biophys. Res. Comm. *80*, 1013 (1978)
58. Clarkson, B., Baserga, F. (eds.): Control of proliferation of animal cells. Cold Spring Harbor Conference on Cell Proliferation, Vol. I, Section I. Cold Spring Harbor, New York: Cold Spring Harbor Laboratories 1974
59. Rein, A., Rubin, H.: Exp. Cell Res. *49*, 666 (1968)
60. Fedoroff, S.: J. Nat. Cancer Inst. *38*, 607 (1967)
61. Melnick, J.L. In: Conference on the application of vaccines against viral rickettsial and bacterial diseases of man. PAHO publication no. 226, p. 171. Washington: PAHO 1971
62. van Wezel, A.L.: Proceedings of the First Meeting of the European Society for Animal Cell Technology, Amsterdam 1976. Published by Rijks Instituut Voor de Volksgezondheid, Bilthoven, Netherlands
63. Moussa, A.A.M., Stouraitis, P.: Report of the meeting of the Research Group of the Standing Technical Committee of the European Commission for the Control of FMD, Brescia/Padua, p. 55. September 1975
64. Jacobs, J.P.: Personal communication
65. Jacobs, J.P.: In: Developments in Biological Standardisation, 42, p. 13. Basel: S. Karger 1979
66. Adams, A., Lidin, B., Strander, H., Cantell, K.: J. Gen. Virol. *28*, 219 (1975)
67. Clarke, J.B., Spier, R.E.: Developments in biological standardisation *35*, p. 61. Basel: S. Karger 1977
68. Author's unpublished observations
69. Montagnier, L., Lage Davila, A.: In: Developments in Biological Standardisation, p. 27. Basel: S. Karger 1979
70. Warren, L., Buck, C.A., Tyszynski, P.: In: Developments in Biological Standardisation, 42, p. 177. Basel: S. Karger 1979
71. Dietrich, C.P., Montes de Oca, H.: Biochem. Biophys. Res. Comm. *80*, 805 (1978)
72. Brown, M., Kiehn, D.: Proc. Natl. Acad. Sci. *74*, 2874 (1977)
73. Mazia, D.: Scient. Amer. *230*, 54 (1974)
74. Folkman, J., Moscona, A.: Nature (Lond.) *273*, 345 (1978)
75. Maroudas, N.G.: Exp. Cell Res. *74*, 337 (1972)
76. McAleer, W.J., Schlabach, A.J., Spier, R.E.: British Patent 1,509,826 (1978)
77. Nordling, S.: Acta Pathol. Microbiol. Scand. Suppl. 912 (1967)
78. Whiteside, J.P., Whiting, B.R., Spier, R.E.: In: Developments in Biological Standardisation, 42, p. 113. Basel: S. Karger 1979
79. Maroudas, N.G. In: New techniques in biophysics and cell biology. Pain, R.H., Smith, B.J. (eds.), Vol. 1, p. 67. New York: Wiley 1973
80. Wako Junyaku Kogyo, Japanese Patent 77035757
81. Author's observations
82. Author's unpublished observations
83. Fontages, R., Beaudry, A., Matot.: Biotechnol. Bioeng. Symp. *4*, 859 (1974)

84. Abbott Laboratories, British Patent no. 1,430,605 (1976)
85. Knazek, R.A., Gullino, P.M., Kohler, P.O., Dedrick, R.L.: Science *178*, 65 (1972)
86. Monsato Company, British Patent 1,514,906 (1978)
87. Jensen, M.D., Wallach, D.F.I.T., Lin, P.S.: Exp. Cell. Res. *84*, 271 (1974)
88. House, W., Shearer, M., Maroudas, N.G.: Exp. Cell. Res. *71*, 293 (1972)
89. Leighton, J. In: Tissue culture. Kruse, P.F., Patterson, M.K. (eds.), p. 367. New York: Academic Press 1973
90. van Wezel, A.L. In; Tissue culture. Kruse, P.F., Patterson, M.K. (eds.), p. 372. New York: Academic Press 1973
91. Spier, R.E., Whiteside, J.P.W.: Biotechnol. Bioeng. *18*, 659 (1976)
92. Levine, D.W., Wang, D.I.C., Thilly, W.G. In: Cell culture and its application. Acton, R.T., Lynn, J.D. (eds.), p. 191. New York: Academic Press 1977
93. Levine, D.W., Wong, J.S., Wang, D.I.C., Thilly, W.G.: Somatic Cell Genetics *3*, 149 (1977)
94. Bloomfield, F.J., Dunstan, D.R., Foster, C.L., Serafini-Cessi, F., Marshall, R.D.: Biochem. J. *164*, 41 (1977)
95. Pearlstein, E.: Nature (Lond.) *262*, 497 (1976)
96. Schlessinger, J., Barak, L.S., Hammes, G.C., Yamada, K.M., Pastan, I., Webb, W.W., Elson, E.L.: Proc. Natl. Acad. Sci., USA *74*, 2909 (1977)
97. Yamada, K.M., Yamada, S.S., Pastan, I.: Proc. Natl. Acad. Sci., USA *73*, 1217 (1976)
98. Steck, T.L.: J. Cell Biol. *62*, 1 (1974)
99. Ponder, E. In: The Cell. Brachet, J., Mirsky, A.E. (eds.). Vol. II, p. 1. London: Academic Press 1961
100. Yogeswaran, G., Steinin, R.W., Wherrett, J.R., Murray, R.K.: J. Biol. Chem. *247*, 5146 (1972)
101. Gahmberg, C.G., Hakomoro, S.: Biochem. Biophys. Res. Comm. *59*, 283 (1974)
102. Steck, T.L., Dawson, G.: J. Biol. Chem. *249*, 2135 (1974)
103. Rothman, J.E., Lenard, J.: Science *195*, 743 (1977)
104. Singer, S.J., Nicolson, G.L.: Science *175*, 720 (1972)
105. Nicolson, G.L. In: Control of proliferation of animal cells. Clarkson, B., Baserga, R. (eds.). Cold Spring Harbor Conference on Cell Proliferation, Vol. I, p. 251. Cold Spring Harbor, New York: Cold Spring Harbor Laboratories 1974
106. Fowler, V., Branton, D.: Nature (Lond.) *268*, 23 (1977)
107. Loor, F.: Nature (Lond.) *264*, 272 (1976)
108. de Mey, J., Hoebeke, J., de Brabander, M., Geuens, G., Jonian, M.: Nature (Lond.) *264*, 273 (1976)
109. Fuller, G.M., Brinkley, B.R.: J. Supramol. Struct. *5*, 497 (1976)
110. Goldman, R.D., Yerna, M.-J., Schloss, J.A.: J. Supramol. Struct. *5*, 155 (1976)
111. Lazarides, E.: J. Supramol. Struct. *5*, 531 (1976)
112. Culp, L.A.: J. Supramol. Struct. *5*, 239 (1976)
113. Okada, T.S., Takeichi, M., Yasuda, K., Ueda, M.J.: Advan. Biophys. *6*, 157 (1974)
114. Monahan, J.J. In: Methods in cell biology. Prescott, D.M. (ed.), Vol. 9, p. 105. London: Academic Press 1976
115. Litwin, J.: In: Developments in Biological Standardisation, 42, p. 172. Basel: S. Karger (Discussion of Session 6) 1979
Basel: S. Karger. (Discussion of Session 6)
116. Maroudas, N.G. In: Cell shape and surface architecture. Revel, J.P., Henning, V., Fox, C.F. (eds.), p. 511. New York: A.R. Liss Inc. 1977
117. Levine, D., Thilly, W.E., Wang, D.I.C.: In: Developments in Biological Standardisation 42, p. 159. Basel: S. Karger. (Discussion of Session 6) 1979
118. Rees, D.A., Lloyd, C.W., Thom, D.: Nature (Lond.) *267*, 124 (1977)
119. Polatnick, J., Bachrach, H.L.: Growth *36*, 247 (1972)

120. Panina, G.F.: Proceedings of the First Meeting of the European Society for Animal Cell Technology, Amsterdam 1976. Published by Rijks Instituut voor de Volksgezondheid, Bilthoven, Netherlands
121. Blakebrough, N. In: Biochemical and biological engineering science. Blakebrough, N. (ed.), Vol. 1, p. 25. London: Academic Press 1967
122. Mougeot, H., Preaud, J.M., Rouchause, J., Favre, H., Dubouclard, C.: Developments in biological standardisation *35*, 33. Basel: S. Karger 1977
123. The Wellcome Foundation, U.K.
124. Aiba, S., Humphrey, A.E., Millis, N.F.: Biochem. Engineering, Chapter 7. New York: Academic Press 1965
125. Oraskainen, P., Lundell, R., Laiho, P.: Process Biochem. *11*, 37 & 55 (1976)
126. Oldshue, J.Y.: Biotechnol. Bioeng. *8*, 3 (1966)
127. Telling, R.C., Radlett, P.J.: Appl. Microbiol. *13*, 91 (1970)
128. McAleer, W.J., Spier, R.E., Posch, K.L.: U.S. Patent No. 3,839,155 (1974)
129. Abercrombie, M.: In Vitro *6*, 128 (1970)
130. Stoker, M.G.P., Rubin, H.: Nature (Lond.) *215*, 171 (1967)
131. Baserga, R., Rovera, G., Faber, J.: In Vitro *7*, 80 (1971)
132. Griffiths, J.B.: J. Cell Sci. *8*, 43 (1971)
133. Holley, R.W., Kiernan, J.A.: Proc. Natl. Acad. Sci. *60*, 300 (1968)
134. Holley, R.W., Armour, R., Baldwin, J.H.: Proc. Natl. Acad. Sci. *75*, 339 (1978)
135. Thrash, C.R., Ho, J., Cunningham, D.D.: J. Biol. Chem. *249*, 6099 (1974)
136. Ceccarini, C., Eagle, H.: Proc. Natl. Acad. Sci. *68*, 229 (1971)
137. Mierzejewski, K., Rozengurt, E.: Nature (Lond.) *269*, 155 (1977)
138. Holley, R.W.: Nature (Lond.) *258*, 487 (1975)
139. Kruse, P.F., Miedema, E.: J. Cell Biol. *27*, 273 (1965)
140. Litwin, J.: Proc. Biochem. *6*, 15 (1971)
141. Sterilin, Data Sheet 8, Teddington, Middlesex, U.K.
142. Nunc Multitray Unit, Kamstrup, 4000 Roskilde, Denmark
143. Linbro Rola Cartridge, 681 Dixwell Ave., New Haven, Conn. 06511, USA
144. Levy, H.B., Riley, F.L., Buckler, G.E.: In: The molecular biology of animal viruses, p. 41. New York: Marcel Decker Inc. 1977
145. Solomons, G.L.: Materials and methods in fermentation, p. 39. London: Academic Press 1969
146. Nicklin, P.M., House, W.: Biotechnol. Bioeng. *18*, 723 (1976)
147. Mann, G.F. Bol. Offic. Sanit. Panamer. (English edn.) *6*, 33 (1972)
148. Chemical Week, August 10, 34 (1977)
149. Kontes, Kontes Instrument Group, Spruce St., Vineland, N.J. 08360 USA
150. Telling, R.C., Passingham, R.J., Kitchener, B.L., Hopkinson, D.G.: British Patent No. 1,436, 323 (1976)
151. Spier, R.E.: Report to the Meeting of the Research Group of the Standing Technical Committee of the European Commission for the Control of FMD, p. 86. Brescia/Padua, Sept. 1975. Rome: FAO 1975
152. Gregg, C.T. In: Growth, nutrition, and metabolism of cells in culture. Rothblat, G.H., Cristofolo, V.J. (eds.), Vol. 1, p. 83. New York: Academic Press 1972
153. Sehgal, P.B., Tamm, I., Vilcek, J.: Virology *70*, 532 (1976)
154. Pearlstein, E., Waterfield, M.B.: Biochim. Biophys. Acta. *362*, 1 (1974)
155. Rudiger, H.W. In: Methods in cell biology. Prescott, D.M. (ed.), Vol. 9, p. 13. New York: Academic Press 1975
156. Sherbet, G.V.: The biophysical characterisation of the cell surface. p. 55. London: Academic Press 1978
157. Mann, G.F.: Wellcome Foundation, British Patent no. 1,369,593 (1974)
158. IL 410 – Instrumentation Laboratory, 113 Hartwell Ave., Lexington, Mass. 02173 USA
159. Pye, D.: J. Biol. Standard. *5*, 307 (1977)

160. Jacob, H.E.: In: Methods in microbiology. Norris, J.R., Ribbons, D.W. (eds.), Vol. 2, p. 91. London: Academic Press 1970
161. Wiles, C.C., Smith, V.C.: 62nd Meeting of American Institute of Chemical Engineers, (Reprint 65b) Washington 1969
162. Daniels, W.F., Garcia, L.H., Rosensteel, J.F.: 62nd Meeting of American Institute of Chemical Engineers, (Reprint 65a) Washington 1969
163. Radlett, P.J., Telling, R.C., Whiteside, J.P., Maskell, M.A.: Biotechnol. Bioeng. *14*, 437 (1972)
164. McLimans, W.F., Blumenson, L.E., Tunnah, K.V.: Biotechnol. Bioeng. *10*, 741 (1968)
165. McLimans, W.F. In: Growth, nutrition, and metabolism of cells in culture. Rothblat, G.H., Cristofolo, V.J. (eds.), Vol. 1, p. 137. London: Academic Press 1972
166. Max-Planck-Gesellschaft zur Förderung der Wissenschaften, E.V. British Patent Nos. 1,389,411 and 1,389, 412 (1975)
167. Kjaeurgaard, L.: Biotechnol. Bioeng. *18*, 285 (1976)
168. Kok, R.: Biotechnol. Bioeng. *18*, 729 (1976)
169. Whiteside, J.P., Spier, R.E.: Developments in biological standardization *35*, p. 67. Basel: S. Karger 1977
170. Beecham Group Limited, Belgian Patent no. 842,002 (1976)
171. Parisius, J.L.E.W., Cucakovitch, N.B., Macmorine, H.M.G.: British Patent no. 1,358,321 (1974)
172. Santero, G.G.: Biotechnol. Bioeng. *14*, 753 (1972)
173. Rewo Laboratory Equipment, 205 Maplehurst Ave., Willowdale, Ontario, Canada
174. House, W. In: Tissue culture. Kruse, P.F., Patterson, M.K. (eds.), p. 338. London: Academic Press 1973
175. Miller, W.J., Spier, R.E., McAleer, W.J.: U.S. Patent no. 3,959,074 (1976)
176. Spier, R.E.: Biotechnol. Bioeng. *19*, 929 (1977)
177. Bradish, C.J., Jowett, R., Kirkham, J.B.: J. Gen. Microbiol. *35*, 27 (1964)
178. Crowther, J.: Developments in biological standardisation *35*, p. 185. Basel: S. Karger 1977
179. Spier, R.W., Whiteside, J.P., Bolt, K.: Biotechnol. Bioeng. *19*, 1735 (1977)
180. Hurni, W.M., McAleer, W.J., Hilleman, M.R.: British Patent no. 1,498,354
181. Munder, P.G., Modolell, M., Wallach, D.F.H.: FEBS Letters *15*, 191 (1971)
182. Radlett, P.J.: Personal Communication
183. Thilly, W.G.: Personal communication
184. Passarge, E., Rudiger, H.W., Wohler, W.: German Patent no. 2,300,567 (1974)
185. Robineaux, R., Lorans, G., Beaure d'Augères, C.: Rev. Europ. Etudes Clin. Biol. *15*, 1066 (1970)
186. Weiss, R.E., Schleicher, J.B.: Biotechnol. Bioeng. *10*, 601 (1968)
187. Litwin, J.: Proceedings of the First Meeting of the European Society for Animal Cell Technology, Amsterdam, 1976. Published by Rijks Instituut Voor de Volksgezondheid, Bilthoven, Netherlands (Paper 2)
188. Fermentation Design, 726 no. Graham St., Allentown, Pa. 18103 USA
189. Girard, H.C., Sutcu, M., Erdem, H., Gurhan, A.: In: Developments in Biological Standardisation 12, p. 127. Basel: S. Karger 1979
190. Knight, E.: Appl. Environ. Microbiol. *33*, 666 (1977)
191. Whiteside, J.P., Spier, R.E., Radlett, P.J.: Unpublished
192. Pagano, J.F., Valenta, J.R.: United States Patent no. 3,740,321 (1973)
193. Ratner, P.L., Cleary, M.L., James, E.: J. Virol. *26*, 536 (1978)
194. Nunc Multitray Unit, Kamstrup, 4000 Roskilde, Denmark, and author's unpublished observations
195. Institut Pasteur, Paris, France. Dutch Patent no. 7,010,119 (1971)
196. Skoda, R., Pakos, V., Johansson, A., Hormann, A., Spath, O.: In: Developments in Biological Standardisation 42, p. 121. Basel: S. Karger 1979

197. French Patent no. 2,352,883
198. Mann, G., Macías, J.: Bull. Pan Amer. Hlth. Org. *10*, 205 (1976)
199. Instrumentation Laboratories Inc.: British Patents 1,490,586 & 1,496,049 (1975)
200. Taylor, W., Evans, V.J.: Biotechnol. Bioeng. *17*, 1847 (1975)
201. Pharmacia, Pharmacia Fine Chemicals AB, Box 175, S 75104, Uppsala 1, Sweden
202. Wohler, A., Rudiger, H.W., Passage, E.: Exp. Cell. Res. *74*, 571 (1972)
203. van Hemert, P.A., Kilburn, D.G., van Wezel, A.L.: Biotechnol. Bioeng. *11*, 875 (1969)
204. Horng, C., McLimans, W.: Biotechnol. Bioeng. *17*, 713 (1975)
205. Spier, R.E.: In: Developments in Biological Standardisation, 42, p. 172. Basel: S. Karger (Discussion of Session 6) 1979
206. Flow Laboratories (Irvine, Scotland)
207. Spier, R.E.: Unpublished observations
208. Meignier, B.: In: Developments in Biological Standardisation, 42, p. 141. Basel: S. Karger 1979
209. van Wezel, A.L.: In: Developments in Biological Standardisation, 42, p. 171. Basel: S. Karger (Discussion of Session 6) 1979
210. Scientific American *238*, 80 (1978)
211. Grinnell, F.: Exp. Cell Res. *97*, 265 (1976)

AUTHOR			ISBN No.
TITLE advances in Biochemical Engineering, Vol. 14, 1980.			EDITION
PUBLISHER Springer-Verlag.	DATE OF PUBLICATION		PRICE £21-08
REQUESTED BY	DEPARTMENT		TICK FOR APPROVAL
ORDER No. 6738	No. of COPIES	SUPPLIER U.B.	DATE ORDER DESPATCHED
REPORTS 27.2.80. BH		NOTES	

R673

Advances in Biochemical Engineering

Editors: T. K. Ghose, A. Fiechter, N. Blakebrough
Managing Editor: A. Fiechter

Volume 10

Immobilized Enzymes I

1978. 48 figures, 14 tables. VII, 177 pages
ISBN 3-540-08975-6

Contents:
W. H. Pitcher, Jr., **Design and Operation of Immobilized Enzyme Reactors.** – S. A. Barker, P. J. Somers, **Biotechnology of Immobilized Multienzyme Systems.** – R. A. Messing, **Carriers for Immobilized Biological Active Systems.** – P. Brodelius, **Industrial Applications of Immobilized Biocatalysts.** – B. Solomon, **Starch Hydrolysis by Immobilized Enzymes. Industrial Applications.**

Volume 11

Microbiology, Theory and Application

1979. 76 figures, 35 tables. V, 180 pages
ISBN 3-540-08990-X

Contents:
D. Ramkrishna, **Statistical Models of Cell Populations.** – S. Nagai, **Mass and Energy Balances for Microbial Growth Kinetics.** – J. M. Scharer, M. Moo-Young, **Methane Generation by Anaerobic Digestion of Cellulose-Containing Wastes.** – B. Metz, N. W. F. Kossen, J. C. Van Suijdam, **The Rheology of Mould Suspensions.** – M. Zlokarnik, **Scale-up of Surface Aerators for Waste Water Treatment.**

Volume 12

Immobilized Enzymes II

1979. 99 figures, 64 tables. V, 253 pages
ISBN 3-540-09262-5

Contents:
H. M. Koplove, C. L. Cooney, **Enzyme Production During Transient Growth.** – R. D. Schmid, **Stabilized Soluble Enzymes.** – S. S. Wang, C.-K. King, **The Use of Coenzymes in Biochemical Reactors.** – C. Wandrey, E. Flaschel, **Process Development and Economic Aspects in Enzyme Engineering. Acylase L-methionine System.** – D. J. Graves, Y.-T. Wu, **The Rational Design of Affinity Chromatography Separation Processes.**

Volume 13

Mass Transfer and Process Control

1979. 134 figures. Approx. 230 pages
ISBN 3-540-09468-7

Contents:
W. Hampel, **Application of Microcomputers in the Study of Microbial Processes.** – Y. H. Lee, G. T. Tsao, **Dissolved Oxygen Electrodes.** – H. Brauer, **Power Consumption in Aerated Stirred Tank Reactor Systems.** – H. Blenke, **Loop Reactors.**

Springer-Verlag
Berlin
Heidelberg
New York

Structure and Bonding

Editors: J. D. Dunitz, J. B. Goodenough, P. Hemmerich, J. A. Ibers, C. K. Jørgensen, J. B. Neilands, D. Reinen, R. J. P. Williams

Springer-Verlag
Berlin
Heidelberg
New York

Volume 20

Biochemistry

1974. 57 figures. IV, 167 pages
ISBN 3-540-07053-2

Contents:

A. S. Mildvan, C. M. Grisham: The Role of Divalent Cations in the Mechanism of Enzyme Catalyzed Phosphoryl and Nucleotidyl Transfer Reactions. – *H. P. C. Hogenkamp, G. N. Sando:* The Enzymatic Reduction of Ribonucleotides. – *W. T. Oosterhuis:* The Electronic State of Iron in Some Natural Iron Compounds: Determination by Mössbauer and ESR Spectroscopy. – *A. Trautwein:* Mössbauer Spectroscopy on Heme Proteins.

Volume 23

Biochemistry

1975. 50 figures. IV, 193 pages
ISBN 3-540-07332-9

Contents:

J. A. Fee: Copper Proteins – Systems Containing the "Blue" Copper Center. – *M. F. Dunn:* Mechanisms of Zinc Ion Catalysis in Small Molecules and Enzymes. – *W. Schneider:* Kinetics and Mechanism of Metalloporphyrin Formation. – *M. Orchin, D. M. Bollinger:* Hydrogen-Deuterium Exchange in Aromatic Compounds.

Volume 29

Biochemistry

1976. 51 figures, 48 tables. IV, 219 pages
ISBN 3-540-07886-X

Contents:

W. G. Zumft: The Molecular Basis of Biological Dinitrogen Fixation. – *J. J. R. Frausto da Silva, R. J. P. Williams:* The Uptake of Elements by Biological Systems. – *A. M. Cheh, J. B. Neilands:* The δ-Aminoevulinate Dehydratases: Molecular and Environmental Properties. – *P. J. Sadler:* The Biological Chemistry of Gold. A Metallo-Drug and Heavy-Atom Label with Variable Valency.